Improving the Technical Requirements Development Process for Weapon Systems

A Systems-Based Approach for Managers

LAUREN A. MAYER, WILLIAM SHELTON, CHRISTIAN JOHNSON,
DANIEL ADDUCCHIO, RAZA KHAN, SUZANNE GENC,
DANIELLE C. TARRAF, NAHOM M. BEYENE

Prepared for the Department of the Air Force
Approved for public release; distribution unlimited

For more information on this publication, visit **www.rand.org/t/RRA997-1**.

Published by the RAND Corporation, Santa Monica, Calif.

Library of Congress Cataloging-in-Publication Data is available for this publication.

ISBN: 978-1-9774-0978-2

Cover: Aliaksandr/AdobeStock, U.S. Air Force.

About This Report

In preparing a request for proposals for a weapon system acquisition, the U.S. Department of the Air Force (DAF) develops a set of technical requirements for the system's design. These requirements dictate the criteria for selecting the prime contractor and for baseline contract negotiations between the government and contractor. These requirements are intended to ensure that the acquired weapon system provides the needed warfighter capability within budget and schedule constraints. However, oversights during this early stage can lead to acquisition cost and schedule overruns—for instance, because of design rework and additional testing—and negative operational outcomes, such as reduced mission effectiveness, when oversights go undiscovered during acquisition. Indeed, a number of U.S. Department of Defense (DoD) programs have seen cost or schedule overruns that can, at least partially, be traced back to errors caused by oversights during development of technical requirements.

To help the DAF improve its process for developing technical requirements for weapon systems, RAND Project AIR FORCE was asked to develop an approach informed by systems-based methods and tools, such as those used in systems engineering, and by one specific emerging hazard-analysis tool: systems-theoretic process analysis (STPA). While STPA has been traditionally used to identify safety hazards in systems, its scope has recently expanded to identifying system hazards that may be associated with mission failure. The approach described in this report is not meant to reproduce or replace the numerous systems engineering handbooks that provide guidance on developing requirements. Instead, it should provide guidance and awareness for those managing and contributing to the DAF's technical requirements development process.

This report should be most useful to program managers in the DAF, who must provide oversight for technical requirements activities. However, other acquisition stakeholders, including the operational, maintenance, and test communities, may also find the report useful, providing awareness of opportunities to contribute and engage in technical requirements development. Furthermore, the report may be more generally useful to organizations exploring the use of STPA for technical requirements development. Finally, other services' program managers and acquisition stakeholders may find value in the approach set out in this report because much of that approach may be applicable to other DoD components.

The research reported here was commissioned by the Director for Global Reach Programs, Office of the Assistant Secretary of the Air Force for Acquisition, Technology and Logistics, Brig Gen Mark R. August and conducted within the Resource Management Program of RAND Project AIR FORCE as part of a fiscal year 2021 project, "Early Identification of Weapon System Acquisition Issues—Application of a Systems Theoretic Approach to the Requirements Process."

RAND Project AIR FORCE

RAND Project AIR FORCE (PAF), a division of the RAND Corporation, is the Department of the Air Force's (DAF's) federally funded research and development center for studies and analyses, supporting both the United States Air Force and the United States Space Force. PAF provides the DAF with independent analyses of policy alternatives affecting the development, employment, combat readiness, and support of current and future air, space, and cyber forces. Research is conducted in four programs: Strategy and Doctrine; Force Modernization and Employment; Resource Management; and Workforce, Development, and Health. The research reported here was prepared under contract FA7014-16-D-1000.

Additional information about PAF is available on our website:
www.rand.org/paf/

This report documents work originally shared with the DAF on August 5, 2021. The draft report, issued on September 22, 2021, was reviewed by formal peer reviewers and DAF subject-matter experts.

Acknowledgments

We thank Brig Gen Mark R. August for sponsoring this work and supporting its execution. We also thank Lt Col Sarah Summers for her guidance and insights throughout this research. Finally, we thank Maj Nathaniel Hagood and Maj Nathan Kopay for their assistance in carrying out this research. Many others in the DAF, too numerous to mention by name, shared their insights with us and participated in discussions and case study research. Additional discussions with individuals from the Georgia Institute of Technology, Massachusetts Institute of Technology, Naval Postgraduate School, Ohio State University, Sandia National Laboratories, U.S. Government Accountability Office, U.S. Navy Air Systems Command, and University of Virginia provided valuable insights.

At the RAND Corporation, we thank Bonnie Triezenberg and George Nacouzi for their analytic reviews and advice; Don Snyder for insightful advice and information throughout the research; Maynard Holliday and N. Peter Whitehead for early contributions to the research; and Silas Dustin, Evan Smith, Rosa Maria Torres, and Francisco Walter for their administrative assistance. We also thank Clifford Whitcomb for his analytic review.

That we received help and insights from those acknowledged here should not be taken to imply that they concur with the views expressed in this report.

Summary

Issue

In preparing a request for proposals to design and produce a weapon system, the U.S. Department of the Air Force (DAF) develops a set of technical requirements—a set of statements or models defining what a system should do and how well it should do it—for the system's design so that the acquired weapon system provides the needed operational capability within budget and schedule constraints. However, oversights during technical requirements development in many U.S. Department of Defense (DoD) programs have resulted in cost or schedule overruns, unsuitable operational performance, or outright cancellation. RAND Project AIR FORCE was asked to develop an approach to help the DAF improve its technical requirements development process. This approach was to be informed by systems-based methods and tools, such as those used in systems engineering, and to include an exploration of the applicability and feasibility of one specific emerging hazard-analysis tool: systems-theoretic process analysis (STPA).

Approach

To inform development of our approach, we sought to answer three research questions:

- How does the DAF currently develop technical requirements and why?
- Can STPA be used in technical requirements development?
- What are best practices for developing technical requirements?

We used a variety of methods to answer these questions, including policy and literature reviews, discussions with DAF stakeholders and systems engineering and STPA subject-matter experts, and case studies of the T-7 and MH-139 programs. Insights from these methods were synthesized and filtered through the institutional roles, responsibilities, and constraints associated with the DAF's technical requirements development to inform our approach and develop associated recommendations.

Key Findings

- Systems engineering activities in the DAF are constrained by the availability of expertise, manpower, training, and guidance. DAF policy and instructions further lack recommended roles and responsibilities for developing technical requirements.
- Possibly as a result, programs' development of technical requirements tends to
  - be ad hoc, often based on the previous experience of available personnel

- rely heavily on technical requirements developed for similar programs, industry standards, market research, and subject-matter expert judgment instead of specific systems engineering methods
    - vary in stakeholder (e.g., operator, maintainer) engagement, with stakeholder roles being dependent on existing relationships and personalities
    - lack defined lines of authority between program offices and stakeholders, leading stakeholders to feel that their involvement in technical requirements development is constrained.
- STPA has the potential to support technical requirements development tasks, but there may not be enough evidence to determine its effectiveness. Other systems engineering tools have proven effectiveness and provide many of the same insights as STPA.
- As with other systems engineering tools, effectively implementing STPA for technical requirements development across DAF programs would require training, further strain personnel time, and entail a nontrivial amount of stakeholder coordination.

Recommendations

- Use the structured, iterative, and tailorable approach for technical requirements development (Figure S.1):
    - Provide it to program managers as tailorable guidance to inform oversight and stakeholder engagement.
    - Update policy to define formal roles to implement it.
    - While implementing the approach may increase early acquisition costs and schedule, the DAF will likely benefit from significant cost and schedule reductions in later phases of acquisition.
- Increase the DAF's organic systems engineering expertise:
    - Work with Defense Acquisition University and the Air Force Institute of Technology to develop and implement training and education.
    - Create a standalone systems engineering field and track personnel who have completed this training, possibly through a specific acquisition workforce series identifier.

Figure S.1. Approach to Technical Requirements Development

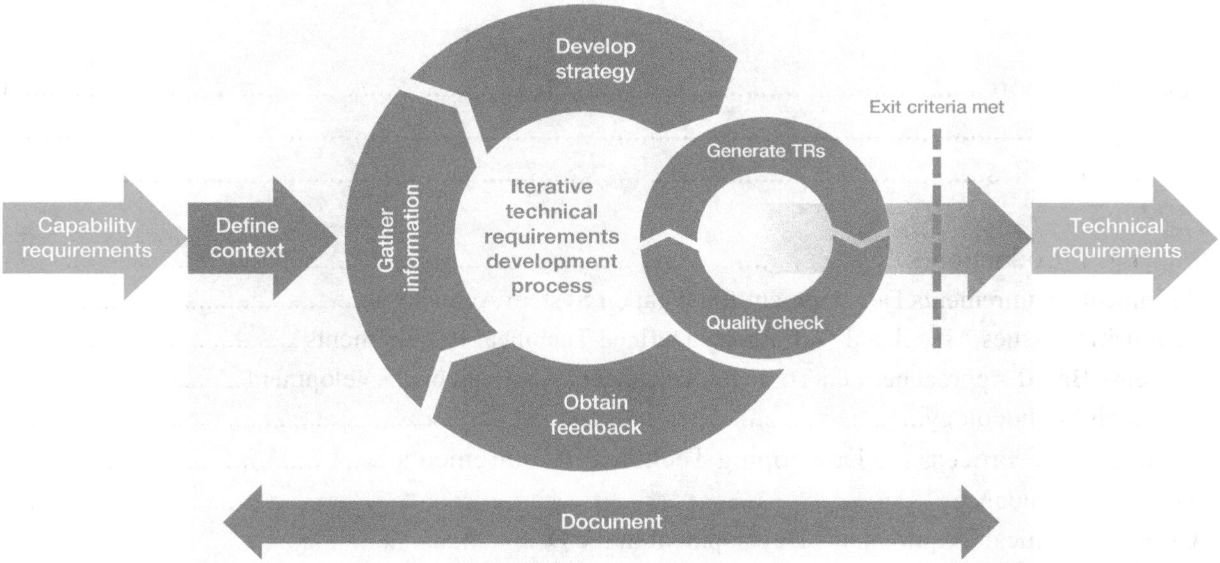

NOTES: The approach aims to convert capability requirements (orange input arrow) to technical requirements (orange output arrow) suitable for inclusion in a request for proposals. It consists of seven elements, shown as blue arrows, with each approach element being comprised of a set of tasks, tools, and stakeholders. TRs = technical requirements.

Contents

Figures and Tables

Figures

Tables

Chapter 1. Introduction

The Department of the Air Force (DAF) uses competitive source selections to contract with industry to design and produce weapon systems. The request for proposals (RFP), integral to the acquisition process, includes a set of technical requirements for the system's design that dictates criteria for selecting the prime contractor and for baseline contract negotiations between the government and contractor.[1] The development of technical requirements for an RFP is one of the final acquisition phases in which the government has full control of and authority over the weapon system's design. Developing technical requirements that are accurate, feasible, and affordable is critical to ensure that DAF weapon systems provide the needed operational capability within budget and schedule constraints.[2] Indeed, a decision to release an RFP for a major development acquisition program (MDAP) must, among other things, "ensure that the program requirements to be proposed against are firm and clearly stated [and] that the risk of committing to development (and eventually production) has been adequately reduced" (Department of Defense [DoD] Instruction [DoDI] 5000.85, 2020).

Still, many weapon systems have experienced cost and schedule overruns because of issues with technical requirements (U.S. Government Accountability Office [GAO], 2015; GAO, 2016; Lorell, Leonard, and Doll, 2015; Lorell, Payne, and Mehta, 2017). While technical requirements are expected to evolve and be further decomposed as a system's design becomes better defined, the research described in this report explores how their development could be improved to reduce unnecessary requirements growth, better control acquisition cost and schedule, and ensure that operational capability needs are met. More specifically, the Assistant Secretary of the Air Force for Acquisition, Technology and Logistics, Global Reach Programs (SAF/AQQ) asked us to develop an approach to improve the DAF's process for developing technical requirements. Given that the use of systems engineering is regarded as a best practice across industry, academia, and government (International Council on Systems Engineering [INCOSE], 2015), our research was informed by the field's many methods, tools, and processes. We were further asked to consider whether an emerging hazard-analysis tool—systems-theoretic process analysis (STPA) (Leveson, 2012)—could be useful to inform technical requirements development. While STPA has been traditionally applied to identify safety hazards in systems, its scope has more

---

[1] *Technical requirements* are a set of statements defining what a system should do and how well it should be done. Those specifying what to do are often termed *functional* requirements. Those specifying how well something is to be done are termed *nonfunctional* or *quality attribute* requirements.

[2] Technical requirements are iteratively refined and expanded after the program contract has been awarded. In this report, we use the term exclusively to refer to the technical requirements used in an RFP prior to contract award.

recently been expanded to identify system hazards that may be associated with mission failure (e.g., Yoo et al., 2021; Scarinci et al., 2019).

This report presents the approach we developed, the research insights that informed it, and broader recommendations to improve the DAF's development of technical requirements. It is worth noting that our approach is not meant to reproduce or replace the numerous systems engineering handbooks that provide guidance on developing requirements. Such guidance will still be useful to acquisition personnel developing technical requirements, and indeed, we frequently refer users to those guides. Instead, our approach is intended to inform management and communication of the process. It should provide foundational awareness for program managers and stakeholders (e.g., implementing command) about relevant objectives, tasks, and tools, allowing better oversight and stakeholder engagement. Furthermore, the approach presented here should be considered to be one that could improve the DAF's technical requirements development. Others may also be effective, but, as we discuss later in this report, any approach that is implemented will likely need to align with the underlying tenets and insights gleaned through this research.

Technical Requirements Development for Weapon System Acquisition

In DoD, technical requirements development is part of the defense acquisition system, described in DoD Directive 5000.01, 2020. DoDI 5000.02, 2020, provides flexibility for DoD components to acquire weapon systems using processes that align with characteristics of the capability being acquired through the Adaptive Acquisition Framework. Regardless of the specific acquisition path used to acquire a weapon system, any time there is a competitive source selection for a development contract that falls under the authority of Federal Acquisition Regulations, the program management office (PMO) will need to develop technical requirements that meet the warfighter's capability needs for an RFP. Therefore, the approach presented in this report could be used in a number of acquisition paths to ensure that technical requirements are appropriately defined for contracting.

The acquisition process for MDAPs is an example of a path involving development of technical requirements for an RFP.[3] Figure 1.1 provides a simplified view of the acquisition process for MDAPs, as outlined in DoDI 5000.85, 2020, and highlights the scope of our research. The MDAP acquisition process uses milestones and decision points to determine whether programs meet the criteria to proceed to the next phase of the acquisition process. Additionally, the warfighter develops capability documents for each phase of the acquisition process: the initial capabilities document (ICD) at Milestone A, the capability development document (CDD) at Milestone B, and the capability production document (CPD) at Milestone C.

[3] This example is based on the major capability acquisition path. The authors understand other paths can be used for programs meeting MDAP requirements, but we use the term *MDAP* here for simplicity.

Figure 1.1. Defense Acquisition Process for MDAPs

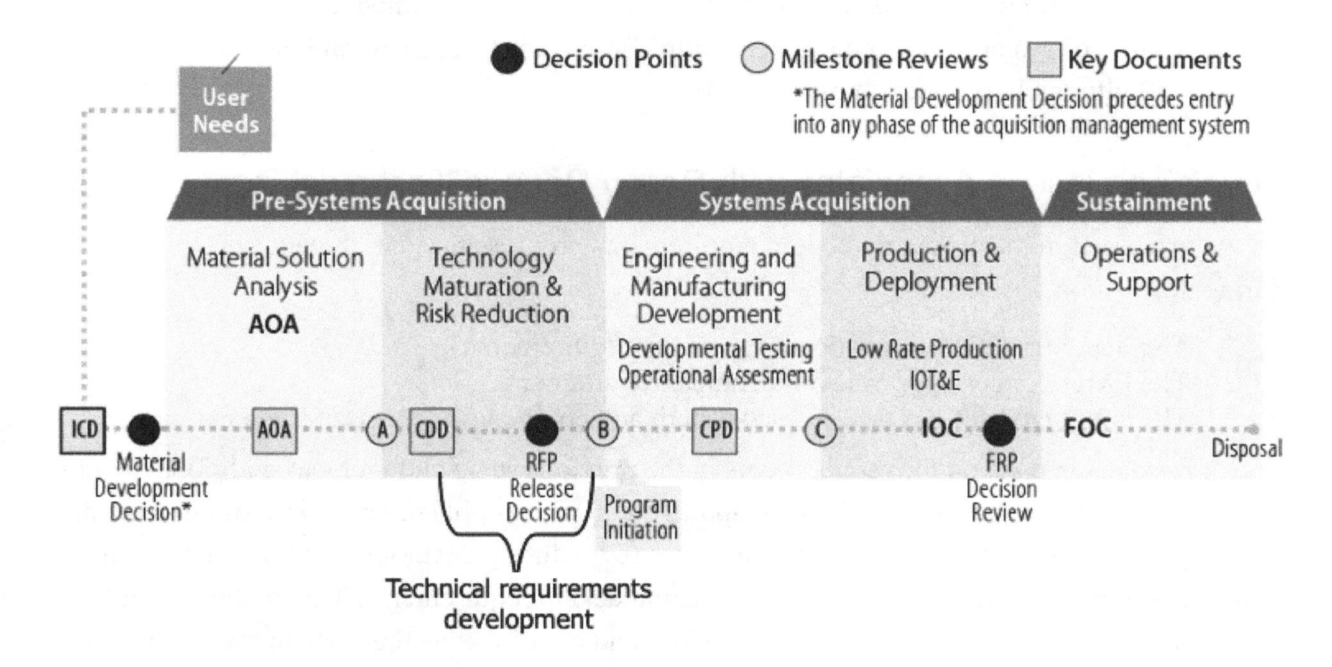

SOURCE: Adapted from Schwartz, 2014.
NOTES: Technical requirements development can occur during other acquisition phases, but this is the research's focal point. AOA = Analysis of Alternatives; IOC = initial operational capability; IOT&E = initial operational test and evaluation; FRP = full-rate production; FOC = full operational capability.

For a major capability acquisition, a number of steps should be accomplished before technical requirements can be developed for an RFP. First, to enter the defense acquisition process, the warfighter (operator) must identify a capability need and perform a capabilities-based assessment (or similar study) showing that a nonmateriel approach cannot address the need. Subsequently, the operator submits a draft ICD to the Joint Requirements Oversight Council (JROC), which validates this need and determines joint capability prioritization. The operator incorporates this feedback, which then must be approved by the JROC as the final ICD.

As Figure 1.1 shows, with the ICD, the milestone decision authority makes the determination about whether a material solution is necessary to meet this need (Milestone Materiel Development Decision). Alternative materiel solutions are explored as part of the AOA, and one is approved as part of the Milestone A review. A CDD is developed by the operator and approved by the JROC to detail the operational capability needs for the anticipated system.

Finally, the CDD is provided to a PMO, which converts the capability needs into the technical requirements that will be used in the RFP, most commonly in a system requirements document (SRD) (the scope of our research efforts).[4] The RFP release initiates a competitive

---

[4] Some MDAPs that have designs heavily based on commercial-off-the-shelf (COTS) or government-off-the-shelf systems may enter the defense acquisition process before Milestone C, in which case, the PMO would develop technical requirements from a CPD.

source selection that results in a prime contractor being chosen for development of the weapon system. With a contract in place, the PMO begins to oversee the detailed design (i.e., the decomposition of system-level requirements into component-level requirements), developmental testing, and ultimately production of the weapon system.

Acquisition Issues Associated with Poorly Defined Technical Requirements

As discussed in the previous section, requirements progress through three generalized stages during acquisition:

1. The operator defines operational capability requirements.
2. The PMO defines system-level technical requirements.
3. The contractor defines detailed subsystem and component-level requirements.

Each stage depends on the requirements in the previous stage and includes additional details that ultimately define the design of the weapon system. Oversights or errors in requirements in one stage may be discovered in subsequent stages (e.g., during developmental or operational testing), requiring additions or revisions to existing design requirements. This requirements evolution can lead to design or production rework and additional testing, which may ultimately increase program acquisition cost and schedule. Of even greater concern, oversights or errors not discovered during development may result in operational weapon systems with design deficiencies. These deficiencies may create safety hazards, reduce mission effectiveness, lead to mission failure, increase costs for operating and maintaining the system, or lead to loss of life.

Preventing (or discovering and mitigating) requirements oversights at their source can therefore reduce adverse acquisition and operational outcomes. A 2015 GAO report analyzed the requirements of 78 MDAPs and found growth in capability requirements to be rare. Instead, the analysis suggested that requirements growth was more directly related to the decomposition of poorly defined system-level technical requirements. This decomposition produced infeasible or unaffordable component-level requirements, resulting in cost growth and other acquisition issues (GAO, 2015). GAO further concluded that programs with infeasible system-level requirements are more susceptible to requirements growth and cost overruns after the contract has been awarded (GAO, 2015).

Two RAND Corporation reports (Lorell, Leonard, and Doll, 2015; Lorell, Payne, and Mehta, 2017) came to similar conclusions by approaching the problem from the opposite direction. The authors compared the key characteristics driving extreme cost growth in six DAF MDAPs with four DAF MDAPs with low cost growth. The analysis revealed that the programs with extreme cost growth were much more likely to have unclear, unstable, and unrealistic requirements before Milestone B. While some of this instability was related to capability requirements, the authors found that high–cost growth programs provided infeasible system-level requirements because they underestimated technical complexity, which led to an unrealistic estimation of the costs associated with achieving such requirements (Lorell, Leonard, and Doll, 2015; Lorell,

Payne, and Mehta, 2017). This research and GAO's analysis clearly suggest infeasible system-level technical requirements as a source of later acquisition issues.

Systems-Based Approaches and Tools for Technical Requirements Development

The previously cited GAO report states that DoD's technical requirements tend to be infeasible because, "DoD often does not perform sufficient up-front requirements analysis via systems engineering on programs" (GAO, 2015, p. 16). A subsequent GAO report asserts that the problem is "directly related to a lack of discipline and rigor in the process of defining and understanding a program's initial requirements," which, it states, can be achieved through detailed systems engineering (GAO, 2016, p. 1).[5] By definition, systems engineering is a top-down, iterative, and interdisciplinary approach for transforming capability requirements (or operator needs) into design requirements with ever more detail at each iteration until a final design is established.[6]

GAO found that early and thorough DoD systems engineering efforts can mitigate the risks and challenges inherent in operational capability requirements (GAO, 2016). It analyzed nine MDAPs to, among other things, illustrate the relationship between systems engineering and program outcomes. Figure 1.2 presents our analysis of GAO's data and findings. GAO, 2016, assessed the level of systems engineering effort that the PMO conducted prior to contract award for each of the nine programs,[7] which could then be compared with the growth in the program's cost and schedule estimates (between the initial and current estimate at time of publication). Our plot of the data suggests a clear relationship between early systems engineering efforts and program outcomes, although the causality of this relationship cannot be determined.[8]

---

[5] An earlier GAO report found that only 11 of 38 reviewed MDAPs held the "key systems engineering event" of preliminary design review prior to the start of the program's contract (GAO, 2013).

[6] The definition of *systems engineering* varies across authoritative guidance. For a more detailed definition, see IEEE 15288 (International Organization for Standardization [ISO], 2015) and Chapter 3 of the *Defense Acquisition Guidebook* (Defense Acquisition University [DAU], 2020).

[7] In our discussions with GAO about these findings, individuals involved with this GAO research stated that the level of systems engineering effort was determined based on the program's ability to achieve a stable allocated baseline (measured against that defined in the critical design review) prior to award of the development contract.

[8] Indeed, GAO defines four program characteristics that can increase the challenges and risks associated with requirements, which could also be driving the relationship presented in Figure 1.2.

Figure 1.2. Relationship Between Early Systems Engineering and Program Outcomes for Nine MDAPs Analyzed by GAO

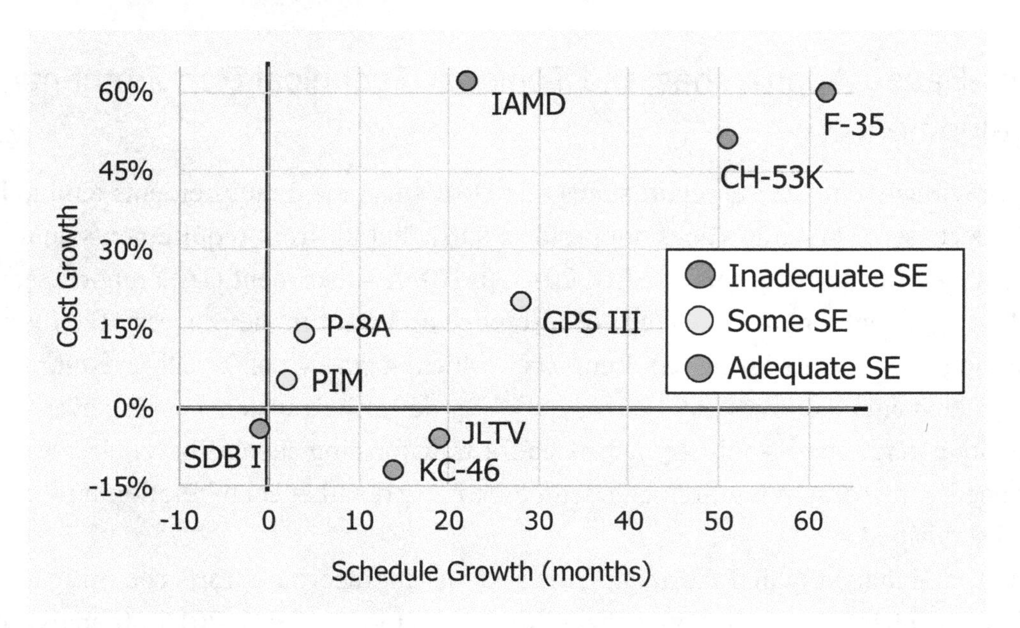

SOURCE: RAND analysis of GAO data (GAO, 2016). Global Positioning System (GPS) III schedule delay estimated from the 2015 Selected Acquisition Report (SAR) (DoD, 2016) as the difference between the GPS III space vehicle 01 Available for Launch SAR baseline production estimate and current estimate.
NOTES: GAO, 2016, provides for the explicit language for and determination of the meanings of *Inadequate*, *Some*, and *Adequate* systems engineering. SDB I = Small Diameter Bomb Increment I; PIM = Paladin Integrated Management; IAMD = Integrated Air and Missile Defense; JLTV = Joint Light Tactical Vehicle; SE = systems engineering.

Systems engineering comprises many methods, tools, and processes (referred to simply as *tools* for the remainder of the report). Systems engineers are trained in how to use these tools and which tool(s) may be most appropriate for the problem at hand. As the field of systems engineering evolves and improves (and as data computation becomes more efficient and smarter), new systems-based tools are developed and discovered that could inform the design process. STPA was originally developed as a hazard-analysis tool to identify system design issues early in the design process that could lead to unacceptable safety losses (Leveson, 2012). By identifying such design issues early in the system design process, STPA may be able to allow such safety hazards to be designed out of the system. More recently, the tool has been used to identify hazards related to cybersecurity (Young and Leveson, 2013; Young and Leveson, 2014) and mission losses (e.g., Yoo et al., 2021; Scarinci et al., 2019). This expansion of the tool's scope has led to questions about whether STPA could be applied during technical requirements development as an alternative to one or more systems engineering tools or as an addition.

Research Methodology

We used a variety of methods to develop a systems-based approach to technical requirements development. Figure 1.3 illustrates the overarching research methodology. We first developed three research questions, each represented in the figure by a box with different color:

- How does the DAF currently develop technical requirements and why? (blue)
- Can STPA be used in technical requirements development? (brown)
- What are best practices for developing technical requirements? (orange)

Figure 1.3. Project Research Methodology

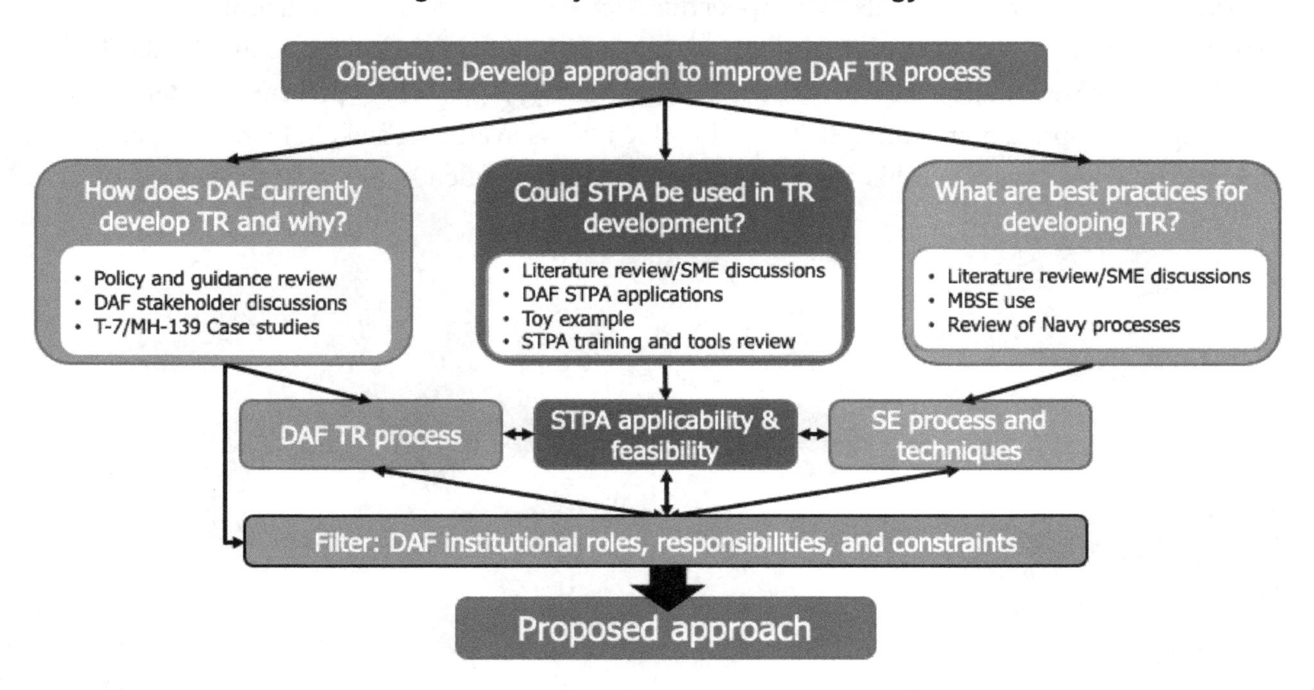

NOTE: MBSE = model-based systems engineering; SME = subject-matter expert.

The methods used to answer each research question are shown below it and include policy and literature reviews, discussions with DAF stakeholders and SMEs, case studies on the technical requirements development processes of the T-7 and MH-139 programs, and a review of U.S. Department of the Navy (DoN) systems engineering processes.[9] These methods revealed insights related to the DAF's technical requirements development process (blue), STPA's applicability and feasibility (orange), and systems engineering process and techniques (green). These insights were synthesized and filtered through the DAF's institutional roles, responsibilities, and constraints associated with technical requirements development (blue) to

---

[9] According to RAND SMEs, the DoN has a strong systems-engineering culture. This, coupled with a question from our sponsor, persuaded the research team to further investigate the DoN's approach to systems engineering.

inform our proposed approach and develop our associated recommendations. Appendix A provides additional details on this research methodology.

The remainder of this report provides our findings and recommendations. Chapter 2 describes the DAF's current process for developing technical requirements, as well as the associated roles, responsibilities, and constraints. Chapter 3 provides an overview of our proposed approach that includes numerous systems engineering processes and tools. Chapter 4 then discusses our findings related to the applicability and feasibility of STPA in relation to the DAF's process and our proposed approach. Chapter 5 presents recommendations to improve the DAF's development of technical requirements across a spectrum of different time frames. Four appendixes provide additional details: Appendix A presents our detailed methodology. Appendixes B and C provide background, results, and insights related to, respectively, the T-7 and MH-139 cases studies and our STPA research. Finally, Appendix D presents the details of our proposed approach, including the tasks involved, tools available, stakeholders required, and operational considerations. This appendix can be used independently of the full report.

Chapter 2. DAF Process for Developing Technical Requirements

Developing an approach to improve the DAF's technical requirements process first requires an understanding of current DAF activities and of the institutional roles, responsibilities, and constraints that underpin these activities. Such insights allowed us to ensure that the approach we developed leveraged lessons from practitioners, improved on areas identified as challenging, and was feasible within the DAF's current institutional infrastructure. In this chapter, we summarize our findings about the DAF's policy and guidance related to technical requirements and associated roles and responsibilities, the generalized process DAF stakeholders use to develop technical requirements, and the constraints or challenges these stakeholders discussed facing when undertaking these activities.

Policy and Guidance

There are few required roles, responsibilities, and activities related specifically to technical requirements development in DAF policy and instructions. Those that are provided are contained in Air Force Policy Directive (AFPD) 63-1, Air Force Instruction (AFI) 63-101/20-101 and Department of the Air Force Pamphlet (DAFPAM) 63-128, 2008, which state the program manager and chief engineer must

- Use a "consistent and rigorous process" for development of technical requirements.
- Identify appropriate design requirements in the RFP by, among other things, integrating capability needs with design considerations to affordably satisfy operator needs and by analyzing and including operator reliability and maintainability requirements.
- Ensure that system requirements include all documented operator requirements (i.e., traceability).
- Coordinate the SRD with the operator prior to release of the final RFP and, more generally, involve the operational, maintenance, and test communities, among others, in systems engineering efforts.

Other roles and responsibilities, as well as activities, are absent from policy to allow PMOs maximum flexibility to tailor the process to a specific situation. While many studies point to the importance of ensuring the feasibility of technical requirements (e.g., Lorell, Leonard, and Doll, 2015; GAO, 2015; Stuckey, Sarkani, and Mazzuchi, 2017), DAF policy explicitly directs only that feasibility be analyzed on capability requirements (AFI 10-601, 2013).

DAF policy and instructions do not provide guidance on processes and techniques for technical requirements development; instead AFI 63-101/20-101 and DAFPAM 63-128, 2008, provide a large number of systems engineering references. Through a structured review of these references, as well as guidance referenced during our discussions with DAF stakeholders, we identified six official or unofficial (e.g., superseded or draft) DoD- or DAF-issued guidance resources relevant to technical requirements development.[1] Through a detailed review of these resources, we found the guidance was generally high level. The resources tend to explain a simplified process for developing technical requirements with limited to no detailed guidance on how to perform the process steps (e.g., tasks or techniques, or roles and responsibilities). Many focus on a set of quality characteristics technical requirements should have that acts almost as a quality assurance checklist. Furthermore, we found the process described in each guidance document to be inconsistent with the others and has limited alignment with the basic direction provided in DAF policy.

While flexibility is indeed important for programs as they develop technical requirements, the lack of authoritative, consistent, and detailed guidance leaves the DAF vulnerable to PMOs adopting unsupported or ineffective approaches that may be more likely to result in adverse acquisition and/or operational outcomes.

Current Technical Requirements Development in the DAF

To determine how the Air Force currently develops technical requirements, including roles and responsibilities, processes, and challenges, we conducted two case studies, on the Grey Wolf (MH-139) and Red Hawk Advanced Pilot Trainer (T-7), as described in Appendix B. We also had discussions with a number of DAF stakeholders, including the PMOs, implementing commands, and test communities associated with each program. We also had discussions with various staff offices or organizations, including the Assistant Secretary of the Air Force for Acquisition, Technology and Logistics, Acquisition Integration Leadership (SAF/AQX); Assistant Secretary of the Air Force for Acquisition, Technology and Logistics, Science Technology and Engineering (SAF/AQR); Air Force Life Cycle Management Center (AFLCMC), Technical Engineering Services Division (AFLCMC/EZ); Air Force Life Cycle Management Center, Plans and Programs Directorate (AFLCMC/XP); the Deputy Chief of Staff for Strategy, Integration and Requirements (AF/A5); the Air Force Institute of Technology (AFIT); and Air Force Materiel Command, Systems Engineering Division (AFMC/ENS). These conversations were supplemented with a review of relevant DAF and DoD policies and instructions.

---

[1] The resources included Chapter 3 of the defense acquisition guide (MIL-HDBK-520A, 2011); a system engineering assessment model (MIL-STD-499B, undated); and two courses in developing technical requirements that were offered to DAF personnel at the time of this research. One of these courses was developed by the Air Force Institute of Technology School of Systems and Logistics (AFIT/LS) (Phipps, 2021) and the other by AFLCMC/EZ (Skujins, 2021). Note that MIL-STD-499B is only a draft standard. It was never approved by DoD, which decided to move toward the use of industry standards for systems engineering guidance. However, our discussions with multiple systems engineering SMEs within and outside the DAF anecdotally point to this draft standard still being used for guidance. For a detailed history of this and other systems engineering standards, see Redshaw, 2010, pp. 95–98.

Technical Requirements Development Roles and Responsibilities

Figure 2.1 illustrates the different organizations with roles and responsibilities for developing technical requirements. At a high level, capability requirements are captured within the capability document (e.g., CDD), developed into technical requirements within the SRD, and further refined in a system specification (SS). Technical requirements development (purple boxes in Figure 2.1) occurs mainly within the PMO, resourced by AFLCMC functional home offices, and industry.[2] The SRD, in conjunction with other relevant documents, is distributed to offerors within industry via an RFP. Offerors' proposals describe various materiel solutions to the government, which the PMO evaluates. Once the PMO awards the development contract to a prime contractor, the two entities negotiate the terms for a statement of work and the system specification.[3]

Figure 2.1. Technical Requirements Development Roles and Responsibilities

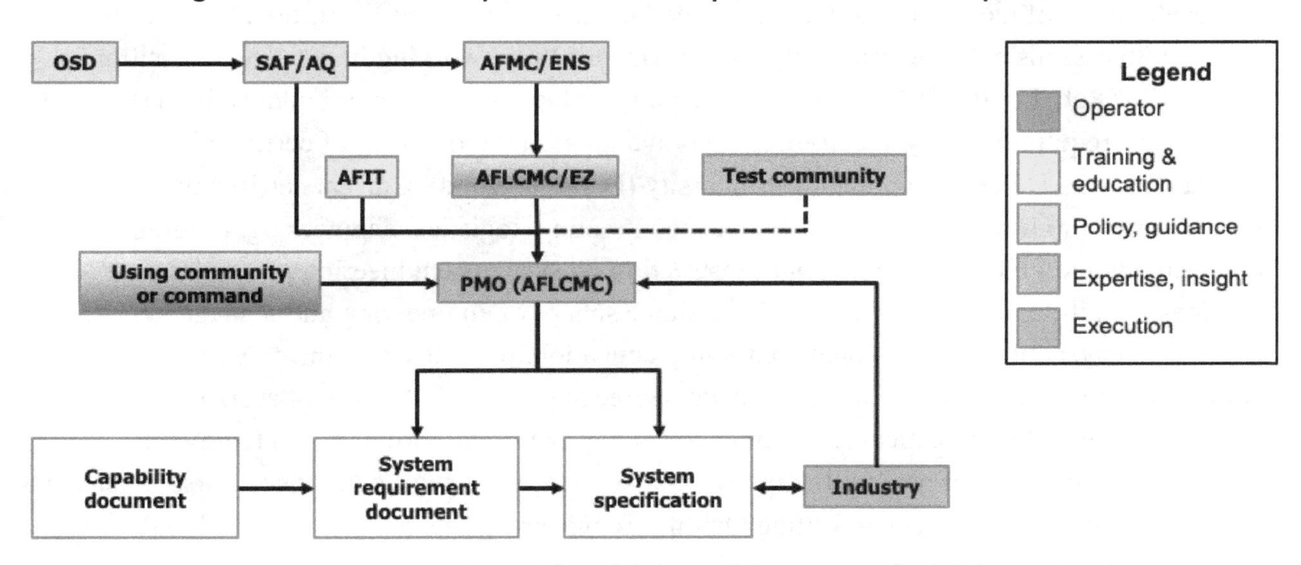

NOTES: PMO is resourced by functional home offices. OSD = Office of the Secretary of Defense.

Several other entities both facilitate and enable development of technical requirements within this process. Expertise and insight are provided by the implementing command, test community and the Technical Engineering Services Division of AFLCMC (AFLCMC/EZ). System operator subject-matter expertise from the implementing command and the operational and developmental test communities provides input on system use and function and details on capability needs.[4]

[2] Industry's primary role is developing a materiel solution meeting the government's technical requirements. However, industry sometimes provides feedback on the system specification.

[3] In certain cases, the PMO may decide to develop a system specification for the RFP (instead of an SRD) to constrain system design negotiations with the prime contractor.

[4] While input from the test communities is useful, it is not always available this early in the acquisition life cycle.

AFLCMC/EZ is available throughout the process to help address challenges or issues encountered.

Policy and guidance (blue boxes in Figure 2.1) are distributed through two routes: the acquisition and management chains of command. Acquisition resides in the office of the Assistant Secretary of the Air Force for Acquisition, Technology and Logistics (SAF/AQ), overseeing Air Force research, development, acquisition, and sustainment activities. SAF/AQX and SAF/AQR develop policy relevant to the technical requirement development process; however, as discussed previously, there is no specific guidance on how to develop technical requirements. The portfolio directorates within SAF/AQ provide specific program guidance for weapon systems within their purview and usually deal with outcomes of implemented policy.[5] The management chain implements policy through AAFMC/ENS, as well as the Air Force Lifecycle Management Center, Engineering Directorate (AFLCMC/EN).

Training and education (yellow boxes in Figure 2.1) are another critical component within the development of technical requirements. The Defense Acquisition Workforce Improvement Act of 1990 established standards and requirements for those working in defense acquisitions (Title XII of Pub. L. 101-510, 1990). The Secretary of Defense further articulates these standards and training requirements by specifying functional areas and indentures of certification levels, as facilitated by the Defense Acquisition University (DAU, undated). Current certification requirements for all functional areas (finance, contracting, logistics, engineering, program management, etc.) include some form of instruction on systems engineering principles and practices. Additionally, AFIT, the DAF's graduate school of engineering and management and institution for technical professional continuing education, provides both military and civilian workforces the opportunity to earn a graduate degree and study in the area of systems engineering, including technical requirements development. Furthermore, AFIT provides continuing education seminars within many acquisition areas. Specifically, one course, Systems and Engineering Management—Writing Quality Requirement Statements (WKSP 0663), is a four-hour seminar on the nuances in writing technical requirements statements, a key element in communicating technical requirements to system developers. AFLCMC/EZ, which plays a dual role in technical requirements development, also reviews guidance and develops training. Specifically, the System Requirements Document Preparation (EZS-107) course covers technical requirements development presented and is provided quarterly.

Technical Requirements Development Process

Our discussions with the various entities highlighted in the previous section suggest that the DAF's technical requirements development follows a generalized process (see Figure 2.2). First, on receipt of the capability document, AFLCMC leadership assembles a team of SMEs as the

[5] The portfolio directorates are Special Programs (SAF/AQL), Global Reach (SAF/AQQ), Global Power (SAF/AQP), Information Dominance (SAF/AQI), and Space Programs (SAF/AQS).

initial technical cadre for the PMO. This initial technical cadre is typically small (e.g., fewer than a dozen members), is drawn from other ongoing AFLCMC activities, and has expertise in areas initially believed to be pertinent to the upcoming materiel acquisition program. The team guides and integrates SMEs, operators, and industry through technical requirements development.

Figure 2.2. Overview of the DAF Technical Requirements Development Process

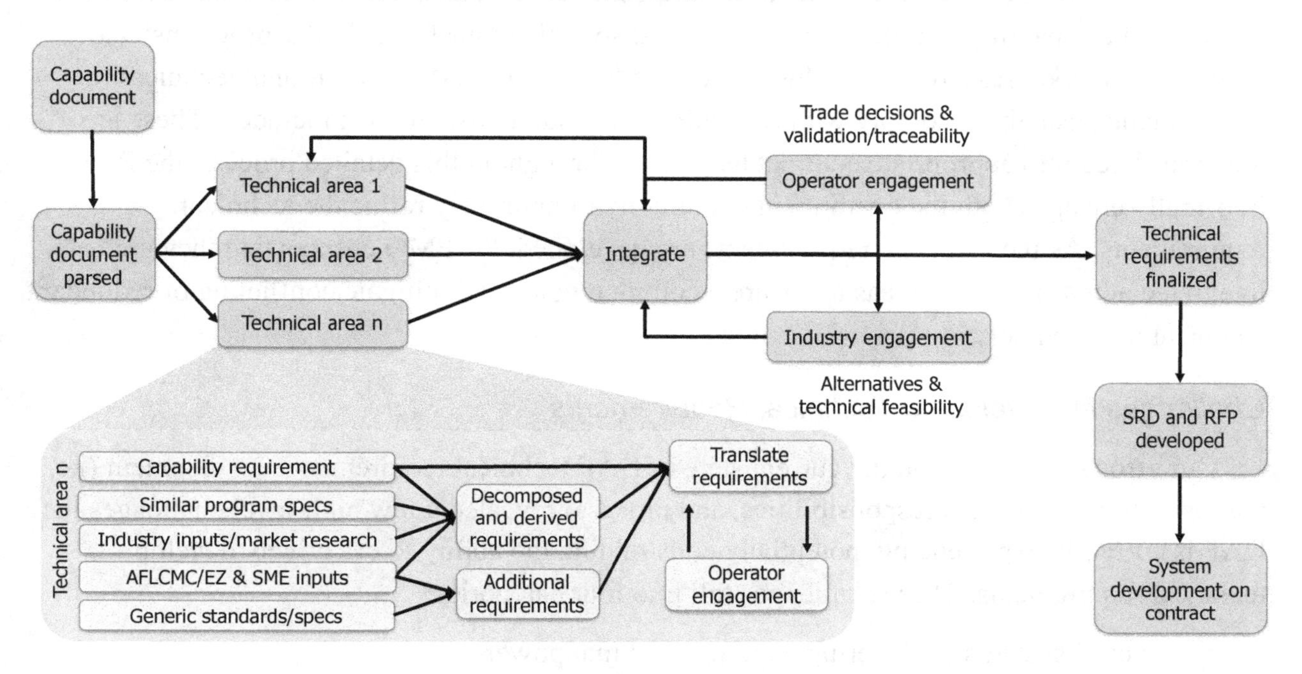

At a high level, the process for developing these technical requirements is iterative to facilitate defining a system that addresses operator requirements within established cost and schedule constraints. The team parses the capability needs into underlying technical areas by initially focusing on key performance parameters and key system attributes. Initial technical requirements from these areas are defined, integrated, and iteratively refined through operator and industry engagement. Industry is leveraged through market research, requests for information (RFIs), and draft RFPs. RFIs are used to learn about industry's capabilities in a particular technology or identify potential solutions to the operational need. Draft RFPs allow the team to obtain feedback on the feasibility and clarity of the government's requirements. Operator engagement occurs both formally and informally. Drafts of technical requirements are shared with the operational community to validate whether they meet the intent of the capability requirement. A review of system design trade-offs between cost/schedule and performance with the operators determines which trade-offs are acceptable for driving the final technical requirements. Through each iteration, the PMO and operators iterate the technical requirements until reaching agreement. The AFLCMC/EN "home office" is also available for consultation or additional expertise. Following the refinements with the using commands and industry, the PMO can finalize the technical requirements and incorporate them into official documents, such as the SRD and RFP.

At the more detailed level, the team and relevant SMEs define technical requirements for each technical area using multiple mechanisms based on experience with previous programs. Some requirements may be copied from the capability document, while others may be further decomposed to a lower level of detail. Many technical requirements are also derived from the original capability requirements or their decompositions. Several sources inform this derivation, including the capability requirements themselves; specifications from similar programs; and inputs from market research and industry, AFLCMC/EZ, and SMEs. Additional technical requirements, usually to comply with regulations and standards, are then included. These are then translated into appropriate contract language. Throughout this detailed process, the PMO informally engages with the operational community to iteratively refine the technical requirements. As the technical requirements are developed, the PMO ensures that they are integrated across technical areas to ensure a common voice and mitigate conflicting or redundant technical requirements.

Challenges in Developing Technical Requirements

Our efforts to understand the current state of DAF technical requirements development (i.e., the institutional roles and responsibilities, and process) revealed a few high-level challenges the DAF is currently experiencing, potentially constraining the ability to effectively develop technical requirements. These challenges fall into four categories:

- level of systems engineering expertise and manpower
- technical requirements development process
- amount and type of stakeholder involvement
- defined lines of authority.

Level of Systems Engineering Expertise and Manpower

We found that systems engineering activities in the DAF are constrained by the availability of expertise, manpower, training, and guidance. During our SME discussions, we learned that organic systems engineering resources significantly decreased in the 1990s.[6] As systems engineering resources became scarcer and as leadership had to manage their use, systems engineering SMEs focused on high-profile programs. DAF stakeholders told us that this resource

[6] After the end of the Cold War, DoD significantly reduced defense personnel, including its acquisition workforce. While the workforce decreased, the number of MDAPs increased. At the same time, DoD policy shifted responsibility for systems engineering activities to the prime contractor developing the weapon system (GAO, 2011). Together, these conditions specifically resulted in workforce cuts to the DoD systems engineering career field (GAO, 2011) and, for the DAF, a lack of experienced systems engineers for early systems engineering efforts for DAF programs (National Research Council, 2008).

allocation is necessary, given that organic manpower is currently at about 60 percent of the requested level (65 percent with contractor support).[7]

AFIT and AFLCMC/EZ do provide training on developing technical requirements to acquisition personnel, but the primary focus is on what language and grammar (translation) to use when writing a requirement, as opposed to how to use systems engineering tools to ensure that technical requirements have the proper attributes (feasibility, affordability, etc.). Also, although two courses are available, they are not mandatory for PMO systems engineering or other interested personnel. Additionally, we learned that PMO personnel may not be aware of the availability of such courses, which are typically advertised through email announcements.

As discussed previously, the guidance available for the PMO to learn and/or follow systems engineering processes and use relevant tools is uncoordinated and inconsistent, with no single authoritative source. The six guidance documents relevant to technical requirements development that we identified summarize the process and do not provide details about how to conduct it.

Technical Requirements Development Process

Given the lack of available systems engineering expertise, training, and guidance and the high demand for this low-density resource, PMOs had to find workarounds to meet programs' schedules. As a result, some PMOs used ad hoc processes to develop technical requirements, often based on the previous experience of their personnel. The PMOs leaned heavily on technical requirements developed for similar programs, relevant industry standards, market research, and available SME judgment instead of specific systems engineering tools. This was verified through our PMO conversations. For example, the MH-139 PMO was not staffed with systems engineers until the program was on contract, long after the technical requirements had been developed, and therefore relied on what organic resources they had.

During the technical requirements process, PMOs infrequently reached out to AFLCMC/EZ to access expertise. While this could supplement a PMO's organic expertise, AFLCMC/EZ lacked the manpower to provide on-call systems engineering expertise to every PMO requesting it.

Amount and Type of Stakeholder Involvement

Developing technical requirements to meet warfighter capability needs involves ongoing communication with the various organizations that interact with the system. This includes operators, but also maintainers and testers. We found that the operator was involved in the technical requirements development process in both our case studies. However, the operators' role was dependent on any existing relationships with and the personalities of the operators and

[7] This estimate was based on AFLCMC/XP's analysis using its Resource Identification Tool.

PMO engineers. Furthermore, we could not find evidence that other stakeholders were involved.[8] Our case study results suggest that this informal, or ad hoc, engagement process was a major challenge to non-PMO stakeholders.

While the MH-139 and T-7 operational communities were involved in the technical requirements development process to varying degrees, we could not find a formal mechanism for either program that ensured stakeholder engagement. In addition, while the operators knew what they wanted to include in the technical requirements, they often felt ill-equipped to communicate effectively with engineers—the difference between "pilot-speak" and "engineer-speak." These factors combined to limit the operators' ability to make informed decisions and communicate their needs to execute their missions effectively.

Defined Lines of Authority

We identified some uncertainty about the lines of authority among the PMOs, operator communities, and test communities. Without a formal mechanism for stakeholder engagement, the operators did not have a process to ensure incorporation of their input. Operational stakeholders for both programs told us that they did not think the PMOs were always including them and listening to their needs. For example, one operational stakeholder told us that operator review of PMO refinements to technical requirements was offered only when the PMO deemed them to affect the system's capability. However, the PMO did not have the operational experience to make such a determination. As a result, operators did not see important changes and did not feel involved in all phases of the technical requirements development.

Summary

This chapter summarizes our findings about the DAF's policy and guidance related to technical requirements and associated roles and responsibilities, the generalized process used by DAF stakeholders to develop technical requirements, and the constraints or challenges these stakeholders discussed facing when undertaking these activities. We found that DAF policy and instructions do not provide guidance on processes and techniques for technical requirements development or on the recommended roles and responsibilities for related activities. Combined with the limited systems engineering expertise and manpower at program startup, these circumstances result in PMO engineers relying heavily on previous experience of available personnel to develop technical requirements and on technical requirements developed for similar programs, industry standards, market research, and SME judgment. While the PMOs for both of our case studies engaged with the operational community, the lack of a defined role for stakeholders meant that there was no systematic method for operational insights to be provided to, or valued by, the PMOs. Furthermore, maintenance and test community engagement was

---

[8] One potential reason the test community was not involved during the technical requirements process in our case studies is that testers are typically not assigned to a program or as part of the PMO until Milestone B or later.

extremely limited or nonexistent. While flexibility is indeed important for programs as they develop technical requirements, our findings suggest that an approach grounded in systems engineering with structured and defined opportunities for stakeholder engagement may improve technical requirements development. The next chapter describes an approach for technical requirements development informed by both the systems engineering literature and the current DAF process.

Chapter 3. Approach for Technical Requirements Development

As discussed in Chapter 1, systems engineering is regarded as a best practice for developing technical requirements (INCOSE, 2015). However, the systems engineering guidance available to the DAF is inconsistent; lacks detail on specific activities, roles and responsibilities; and does not account for the institutional organization and challenges of the DAF. In this chapter, we describe an approach informed by both the systems engineering literature and the current DAF process. This chapter's goal is to provide program managers and other stakeholders a high-level sense of how to successfully develop technical requirements from operational capabilities in a capability document (e.g., CDD, CPD).

Appendix D provides a more detailed description of the approach and presents a number of implementation considerations. The appendix can be used independently of the full report and as tailorable guidance for the DAF's development of technical requirements. It describes each element and task outlined in the approach, including task inputs and outputs, and recommended stakeholder roles and responsibilities. The appendix also briefly describes 12 systems engineering tools, a cross-section of the tools available for operationalizing the approach; maps them to relevant tasks; and provides references for the interested reader.[1]

This chapter and the approach in Appendix D do provide some guidance for systems engineering concepts and tools but are not meant to reproduce or replace the numerous systems engineering handbooks that provide guidance on developing requirements. Such guidance is still useful to acquisition personnel developing technical requirements; indeed, we frequently refer users to these guides. Instead, the approach described in this report should provide a structure to inform oversight and stakeholder engagement.

This chapter begins with an overview of the approach, which we leaned on multiple sources to develop. We then describe some of the underlying tenets that informed our approach, gleaned from themes that emerged from our stakeholder discussions, case studies, and systems engineering guidance review. Finally, we use our approach as a framework to attempt to draw relationships between technical requirements development activities performed in our case

---

[1] One of the tools highlighted in the appendix is MBSE. MBSE is more than just a tool—it is an engineering paradigm that is advancing the field of systems engineering and is increasingly being adopted both inside and outside the DAF. Some of the tasks in our approach (e.g., deriving and analyzing technical requirements) can be automated using MBSE. Others, such as understanding stakeholder needs, translating technical requirements, and obtaining industry feedback, will still need to occur even with the most mature MBSE processes in place. MBSE will undoubtedly change the way the DAF develops technical requirements; as we contend throughout the report, other approaches may also improve the DAF's process. However, this chapter also provides a set of tenets that the DAF can use as nonprescriptive guidance when developing technical requirements. These tenets should remain relevant even as the DAF transitions to MBSE.

studies and the current status of the programs. Throughout the chapter, we reference some tools that may be of value, but defer further discussion of these tools to Appendix D.

Approach Overview

Our proposed approach intends to overcome the various constraints described in Chapter 2 in the context of existing DAF processes and systems engineering best practices and tools. More critically, the proposed approach is not intended to be prescriptive. Instead, the objective is for the approach to remain tailorable to a specific weapon system and capability need and to the circumstances for the problem at hand, while providing sufficient high-level guidance to the program managers overseeing them.

The approach draws heavily on the processes and tools described in numerous systems engineering guides (e.g., DoN, 2004; DAU, 2020; ISO, 2015; NASA, 2020). While many of these guides were informed by the same underlying principles, they diverged over time to suit the organization and/or industry for which they were developed (Redshaw, 2010). Our approach attempts to generalize across these guides to define a set of basic elements the DAF can use to develop technical requirements for its weapon systems. Figure 3.1 depicts this approach and its seven elements (in the dark blue arrows).

Figure 3.1. Approach to Technical Requirements Development

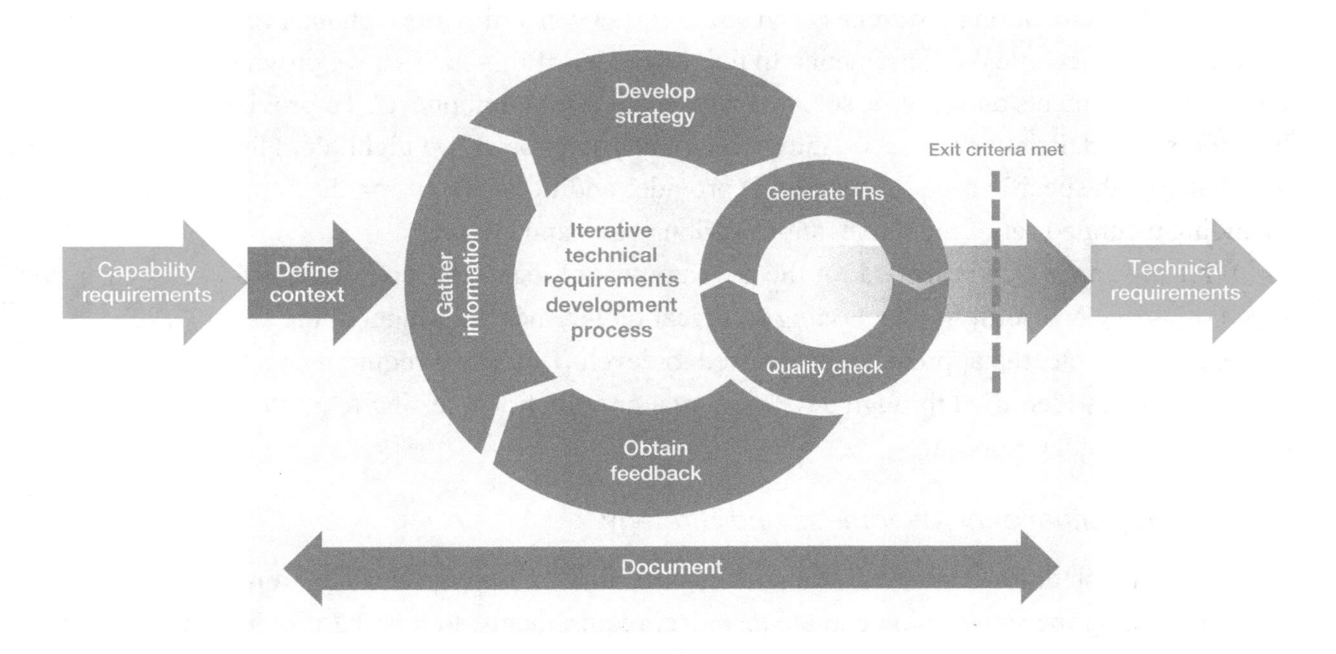

NOTES: The approach aims to convert capability requirements (orange input arrow) to technical requirements (orange output arrow) suitable for inclusion in a request for proposals. It consists of seven elements, shown as blue arrows, with each approach element being comprised of a set of tasks, tools, and stakeholders. TRs = technical requirements.

As shown in the figure, the approach aims to convert capability requirements (orange input arrow) to technical requirements (orange output arrow) suitable for inclusion in an RFP. The approach consists of seven elements, shown as blue arrows, with each approach element comprising a set of tasks, tools, and stakeholders.

The figure highlights a few important underlying concepts. First, the approach is inherently iterative, with multiple feedback loops. Technical requirements should be refined as more information is identified and as decisions are reached. Relatedly, the approach includes a set of exit criteria that need to be objective and measurable and that clearly delineate the point at which it becomes possible to exit the process with a set of quality (e.g., feasible and affordable) technical requirements that meet the desired capabilities.[2]

Second, it is critical to document the process to capture the lessons learned and the decision rationales from beginning to end. This helps ensure traceability of decisions through the various iterations, decision cycles, and potential personnel changes throughout the process. Documentation should not place significant additional burden on staff and stakeholders.

Third, the approach distinguishes between *developing* and *generating* technical requirements. In this report, *developing* technical requirements, refers to the entire process depicted in Figure 3.1, including information gathering, eliciting feedback, and strategic decisionmaking. *Generating* technical requirements refers to the process used to define the technical requirements that will be used in the RFP.

Finally, while iterative, the approach does follow a general sequence through each iteration: (1) establishing situational awareness and strategy, (2) generating the technical requirements, and (3) refining the technical requirements. In the remainder of this section, we provide a brief overview of elements and tasks associated with these three components. To provide context for how tools and stakeholder roles are intertwined in our approach, we highlight a few of each associated with specific tasks. Appendix D provides additional details on the approach elements, including a curated set of tools relevant to various tasks and elements.

While this approach is scoped for the development of technical requirements for an RFP, much of it may be useful for developing technical requirements throughout the acquisition life cycle. For example, the approach may be used to develop technical requirements for additional capability needs identified through developmental or operational testing (e.g., through engineering change proposals).

Establishing Situational Awareness and Strategy

Our approach defines three elements for establishing the appropriate situational awareness and determining the strategy to generate technical requirements. Figure 3.2 provides an overview of the tasks and stakeholders involved in each element highlighted in red. The PMO should first

---

[2] Because these criteria are capability and program specific, this report does not address them. However, they should help ensure the technical requirements are feasible, measurable, and affordable.

define the context surrounding the system, which requires identifying all relevant stakeholders and ensuring the office understands their needs. Each stakeholder should communicate its goals and priorities for the program, including how the system will be used and its interaction with the operational environment. These can be captured using an operational diagram that details system interactions across the system's entire life cycle. For example, an Operational View 1 (OV-1) can help stakeholders communicate this context (DoD Chief Information Officer, 2010).

Figure 3.2. Situational Awareness and Strategy

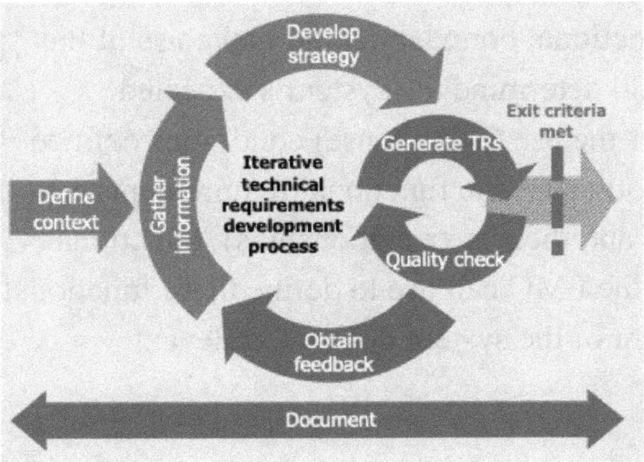

Element	Task	Stakeholders
Define context	Identify and understand stakeholders needs	SPO, operator, maintainer, legacy system operators
	Develop or understand operational diagram	
Gather information	Identify system constraints/considerations	SPO, operator, AFLCMC/EZ, testers, other SPOs
	Identify relevant system information	
Develop strategy	Develop technical requirements generation process	SPO, operator
	Organize and prioritize technical areas	

NOTE: SPO = system program office.

Once the context is understood, the PMO can begin to identify and collect information about the system. System constraints, such as the maturity and availability of the market for potential technology solutions, can be collected using market research (e.g., RFIs) (SD-5, 2008). DoD specifications and regulatory requirements should be identified, and environmental conditions for use cases should be explored. Information related to analogous or legacy systems can also be gathered to understand lessons and pitfalls.

With proper situational awareness, the PMO can work with the operational community to develop a strategy for generating the technical requirements for the system. This should include the processes, milestones, and tools to be used, as well as stakeholder roles, communication feedback mechanisms, and adjudication processes. The strategy should consider program constraints, such as budget, available expertise, and acquisition strategy, and determine how resources, roles, and responsibilities should be allocated accordingly. For example, critical capabilities identified while defining the context may support additional expertise and resources allocated to specific technical areas of system.

For some programs, many tasks in these three elements may be partially performed to inform the development of the capability document (e.g., during developmental planning activities). In those cases, duplication of effort should be minimized, but the PMO and operator should ensure that personnel not involved in these earlier activities (e.g., some members of the PMO's initial cadre) obtain the needed situational awareness to perform later approach tasks and that the strategy is revised for any additional information gathered.

Generating Technical Requirements

The fourth element in the approach encompasses five tasks to generate the technical requirements in a top-down manner (see Figure 3.3). While the PMO should lead these tasks, it is very important to engage the operator and the maintenance and test communities as needed. These tasks should follow the strategy developed in the third element (i.e., Develop strategy) and be heavily informed by the knowledge gained and information gathered to establish situational awareness. For example, the first task to define the functional boundary could make use of the operational diagram and system constraints. This would determine the system's expected behavior and the allowable bounds of that behavior for the use (and misuse) conditions defined in the first element (i.e., Define context). Given this boundary, the functional requirements can be developed (i.e., the desired functions of the system and their interrelationships). Functional flow block diagrams (FFBDs) (DAU, 2001) are tools the PMO can use to define these functional relationships and document and communicate the intent of the system design to relevant stakeholders.

Figure 3.3. Generate Technical Requirements

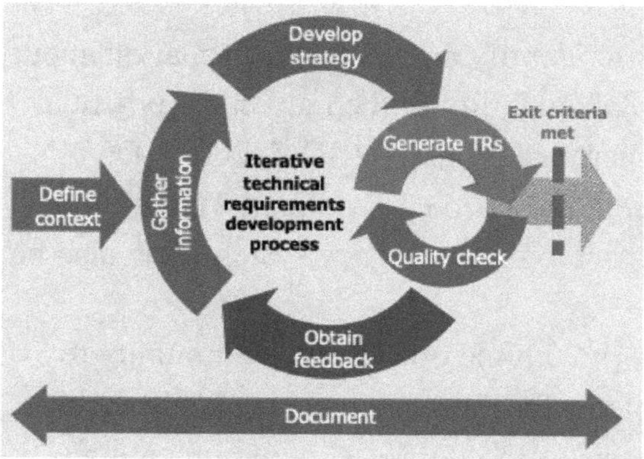

Element	Task	Who
Generate technical requirements	Define functional boundary (expected system behavior and its boundary)	SPO, operator, maintainer, tester, AFLCMC/EZ
	Define functions (what system must do) and decompose as necessary	
	Derive technical requirements (e.g., performance, availability, safety requirements) using appropriate derivation methods	
	Refine technical requirements to incorporate constraints (e.g., DoD specifications)	
	Analyze, trade off and refine requirements (e.g., for affordability, feasibility, allocation)	

For each function, the quality attribute requirements can then be derived.[3] These quality-attribute requirements define the functional conditions under which the system must operate, such as its performance, reliability, maintainability, and safety. Numerous requirement-derivation methods exist, but all require the involvement of relevant stakeholders to ensure that the derived requirements align with stakeholder goals. For example, a facilitated and structured stakeholder workshop, called a quality attribute workshop (QAW) (Barbacci et al., 2003), can be

[3] Sometimes these are referred to as *nonfunctional* requirements. However, this could trivialize their importance so we choose to use *quality attributes* instead.

held to develop the system's quality attributes (e.g., reliability, availability) for a prioritized set of system use scenarios.

After each draft of technical requirements has been generated, they should be refined to incorporate compliance and regulatory requirements, such as DoD specifications and industry standards. Technical requirements for analogous systems can also be considered. Finally, the set of technical requirements must be analyzed for, among other characteristics, affordability, feasibility, and allocation of physical qualities (e.g., power, weight). The PMO can develop a tradespace that captures the sensitivity of these characteristics to changes in technical requirements. Such trades (e.g., worse performance for a more affordable system) should be presented to the operator and other stakeholders to elicit their preferences and to refine the technical requirements accordingly. While not represented in the approach figure, each draft of generated technical requirements will require iterations before the fifth element (i.e., Quality check) in the approach can be undertaken.

Refining Technical Requirements

The approach next includes two opportunities to review the draft set of technical requirements for needed refinements (see Figure 3.4). The first is a quality-assurance assessment. Given that subsets of the technical requirements may be developed in isolation (i.e., within technical areas), they will need to be integrated to identify any redundancies, gaps, or conflicts. Identifying any will likely result in feedback to the previous approach element (i.e., Generating technical requirements) for refinement. Once technical requirements are fully integrated, the PMO can translate them into contract language (e.g., use *shall* statements). A final quality-assurance assessment can be completed using a standardized checklist that ensures that each

Figure 3.4. Refinement of Technical Requirements

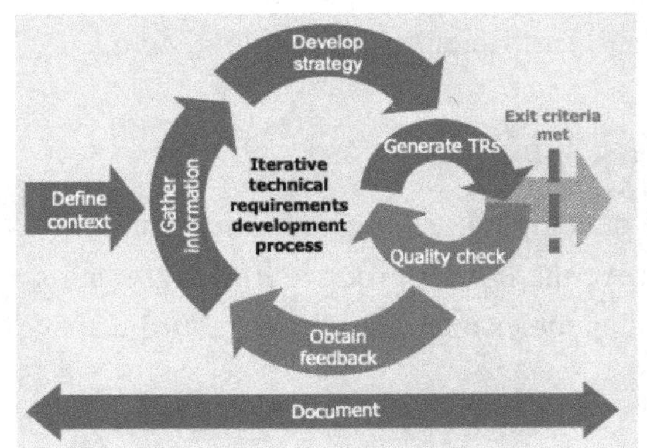

Element	Task	Who
Quality check	Integrate technical requirements (identify gaps, conflicts, etc.)	SPO, operator
	Translate technical requirements (i.e., into contract language)	
	Assess quality of requirements (for language, completeness, traceability etc.)	
Obtain feedback	Stakeholder feedback and adjudication (formal mechanism)	SPO, operator, maintainer, tester, AFLCMC/EZ
	Red team/independent review	
	Industry feedback (e.g., draft RFPs)	

individual requirement and the set as a whole are well-written and meet quality standards.[4] Again, this quality assurance may lead to feedback into the "generating technical requirements" element for refinement. Once the technical requirements have passed the quality check, three tasks lay out formal mechanisms for review from stakeholders outside the PMO. The first task recommends that stakeholders, including the operational, maintenance, and test communities, provide feedback and that feedback be adjudicated according to the process agreed to during the develop strategy element. The second involves red teaming or an independent review. This task can take on many forms but should include individuals already familiar with the system and mission to reduce the need for additional personnel and resources. An organization outside the PMO or operational community, such as AFLCMC/EZ, could oversee the effort. As part of this review, participants should question the technical requirements' underlying assumptions (e.g., feasibility, system use conditions) and review documentation on efforts to generate and analyze requirements. Together, these can help uncover the potential for incorrect, inaccurate, infeasible, or missing technical requirements. One tool that may be useful for this type of review is STPA (Leveson and Thomas, 2018) because it can discover scenarios that could lead to unacceptable losses (e.g., mission, safety, or security failure). The final task in this element includes eliciting and adjudicating feedback from industry, for example, through draft RFPs.

These formal review mechanisms are likely to reveal further questions about the capability needs, system constraints, or expected behavior. Answering such questions may require gathering additional information, resulting in further iterative refinements to the technical requirements.[5] During the quality check of future iterations of this process, the exit criteria developed as part of the technical requirements–generation strategy can begin to be assessed. Once all criteria have been met, the technical requirements can be included in the final RFP.

Adopting an approach like the one outlined above should greatly support minimizing, if not eliminating, oversights and errors in technical requirements. Following this approach may take longer than the existing DAF process, but this would likely be offset by reductions in future cost growth and schedule slips, while producing an operationally sufficient and suitable system.

Underlying Tenets

In the previous section, we defined one approach to developing technical requirements for DAF weapon systems. Other approaches may also improve the DAF's process. Indeed, systems engineering guidebooks developed by other government agencies and industry (e.g., DoN, 2004;

---

[4] Appendix C of the *NASA Systems Engineering Handbook* (NASA, 2020) has a particularly comprehensive checklist.

[5] In some cases, the refinement of technical requirements may call into question the context definition developed at the start of the process. This may result in broader changes to the capability document (or its interpretation) or operational diagram related to the Define Context task. In these cases, the SPO and stakeholders may need to revisit this first task before further refining the technical requirements.

DAU, 2020; ISO, 2015; NASA, 2020) all offer somewhat different approaches. Certain themes emerge as cross-cutting during our reviews of these guidebooks and other relevant literature and many discussions with systems engineering SMEs and DAF stakeholders. These themes may provide a set of tenets that can be used as nonprescriptive guidance for DAF programs as they work to implement the approach presented here or another that better applies to their current circumstances (e.g., available funding, expertise):

- ***Develop a plan.*** The schedule and funding for RFP development is often set by the acquisition strategy and other factors and may not necessarily align with what is needed to develop technical requirements most effectively. While planning will require up-front time and effort, it should also save time and resources in the long run. This includes
  - understanding the programmatic constraints (e.g., funding, expertise) and operator needs, and prioritize resources accordingly
  - defining the strategy for generating technical requirements, including the tasks, tools and stakeholders needed
  - determining how to adjudicate comments and/or disagreements from all stakeholders
  - not underestimating the level of coordination necessary to involve stakeholders, elicit and adjudicate feedback, and refine the technical requirements.

  Early planning will make the process more efficient and effective.
- ***Encourage two-way communication.*** Opportunities for both formal and informal communication between the PMO and all stakeholders are needed. Cast a wide net for stakeholder inclusion. Frequent communication with the implementing command and industry is very important, but feedback from the test and maintenance communities, as well as legacy and similar-system engineers and operators, will greatly inform technical requirements. In the spirit of two-way communication, all stakeholders, including the PMO, should receive as much as they transmit.
- ***Adopt a continuous-learning approach.*** Initial drafts of technical requirements should be conceptualized as a means of obtaining additional information and knowledge. Use these drafts to elicit feedback (e.g., from industry and operators) and better learn about technical requirements feasibility and affordability (e.g., through requirements analysis). As new knowledge is gained, updates to the strategy for generating technical requirements may be necessary, along with updates to the technical requirements themselves.
- ***Exploit existing opportunities for documentation.*** Documentation ensures the processes used and the rationale for technical requirements decisions (e.g., assumptions, simulation results, priorities) are captured for subsequent iterations of technical requirements or injection of new personnel. However, formal documentation can claim the resources sorely needed for other tasks. Fortunately, existing artifacts of technical requirements development can fulfill much of the documentation needs for the process. Documentation may be as simple as establishing a web-based site to store important emails, technical data, and meeting notes.
- ***Define the boundaries for the system and/or mission.*** Every systems engineering guidance document we reviewed and expert we spoke with discussed the importance of

developing a clear understanding of the system and/or mission boundary. How will the system be used? How will it interact with its surroundings and other systems throughout its entire life cycle? How should the system behave and not behave in both expected and unexpected scenarios? Answers to these and similar questions ensure that technical requirements are comprehensive.

- ***Use a top-down approach.*** Generate technical requirements based on the capability requirements provided and iteratively refine the technical requirements to address such considerations as DoD specifications, industry standards, and information from analogous systems. Beginning with a template (e.g., based on analogous systems) constrains the solution space and risks omitting technical requirements needed to best meet the capability. If a template is used, all the assumptions it contains must be questioned.
- ***Make balance a priority.*** Developing technical requirements necessitates balancing many different objectives. Technical requirements must balance operator capability needs against affordability and technical feasibility. They must also balance the level of specificity: over-specifying requirements (e.g., unnecessary decomposition) can constrain the solution set and market interest but under-specifying them can lead to operator capability needs not being met.[6] There will be the need to balance acquisition against sustainment costs, requirements verifiability against traceability, technical feasibility against desired performance, and system usability against survivability, etc. The multiobjective tradespace for requirements should be defined and each objective prioritized with input for relevant stakeholders.

Case Study Retrospection

The approach presented in this chapter and Appendix D was heavily informed by the processes used to develop technical requirements in our case studies. Given that development contracts for both the T-7 and MH-139 programs have been awarded and that developmental testing is underway,[7] we were able to explore the status of each program and attempt to draw relationships to activities undertaken during technical requirements development. Such analysis is speculative, given the lack of a counterfactual case, but allowed us to highlight areas that may have benefited from our approach. In this section, we discuss some examples of how our

---

[6] This concept of specifying requirements only to the level necessary aligns with the "just enough, just in time" concept from lean engineering. Technical requirements should be developed that use just-in-time information and just enough detail to communicate the desired design elements. Every architecture decision that is made will constrain the design space. This means obtaining enough knowledge to understand which technical requirements are essential, as opposed to being nice to have before decisions are made. For more information on this concept, see Leffingwell, 2011.

[7] Both had firm-fixed-price (FFP) contracts, which move all developmental risk to the contractor. While this can be an effective method for controlling costs, it could result in unanticipated contractor behavior as the contractor balances performance against costs. Cost pressures might cause the contractor to view a technical requirement, no matter how well-written, differently from the way government stakeholders do. Therefore, FFP development should be viewed with caution.

approach might have improved acquisition outcomes in these two programs. Appendix B offers more details about the case studies.

Comparing the technical requirements development for the T-7 and MH-139 revealed several differences and commonalities. One primary difference is that the T-7 capability document was developed concurrently with the technical requirements, while those for the MH-139 were developed in the more traditional sequential order. This environment facilitated continual collaboration between the T-7 PMO and the operator, the Air Education Training Command (AETC). Although the MH-139 operator, Air Force Global Strike Command (AFGSC) communicated frequently with the MH-139 PMO, our discussions with stakeholders from both programs suggest that AFGSC had more issues with the capability of the weapon system in development than AETC. The MH-139 PMO did gain some knowledge of the operational context through a set of use cases, although AFGSC claimed that this was insufficient and that this is reflected in the current capability of the system. Our approach continually recommends not just operator engagement but two-way communication to ensure alignment of operator and PMO understanding. The MH-139's operator and test communities both stated that, although technical requirements in the SRD adequately represented capability needs, much of the important language was changed when the PMO worked with the contractor to develop the system specification. Better alignment of operator and PMO understanding or updating the other stakeholders during the process could have helped ensure that this language was retained or the change was justified.

Both programs lacked heavy stakeholder involvement and are experiencing human integration and/or concept of employment (CONEMP) issues. In the case of the T-7, the aircraft ladder, as currently designed, will likely create difficulties for operator egress but meets the written technical requirements. Similarly, the MH-139's gun and seat placement requires additional operator effort when shooting but, again, meets the written technical requirements. Representatives from AETC and AFGSC, respectively, agreed that the requirements are correct but leave room for contractor interpretation. In both cases, with an FFP contract in place, the contractor is incentivized to interpret technical requirements in the most cost-efficient manner, which may not also be the most operationally effective solution. Input from pilot and/or legacy system operators while defining the context and/or during feedback elicitation could have revealed such oversights.

The MH-139 has multiple examples in which a more-thorough set of use cases could have better informed requirements. For example, the technical requirements on the contract specify an "austere landing capability"; AFGSC stated that contractors might not interpret this as the capability to land on an unprepared landing surface (e.g., in a mountainous area). AFGSC told us that, while most operators will land the helicopter on heliports, this austere landing capability is needed under certain circumstances. As another example, we learned that the expected use of the helicopter has evolved such that it will be used more often, which increases the need for maintenance. However, the FFP logistics and parts supply contract has incentivized the contractor to trade maneuverability for sustainability, exploiting underspecificity in the technical requirements. As we discuss in Appendix D, defining the operational context should also involve considering misuse cases to explore how the system should behave under off-nominal conditions. Such an exercise may have revealed the need to include maneuverability in the technical requirements.

As one last example, technical requirements for both systems were developed by starting with technical requirements from legacy systems and refining these based on capability requirements, industry feedback, and other constraints and considerations. Our approach recommends developing requirements from the top down, which requires fully understanding stakeholders' needs, the operational context, and any constraints to generate a set of technical requirements based on the functions the system must perform, with later refinements informed by the legacy system. Using a top-down approach helps ensure that the technical requirements are more likely to be comprehensive. Some of the examples provided here, which demonstrate potential oversights in derived requirements, might have been avoided by using a top-down approach.

Chapter 4. Systems-Theoretic Process Analysis Applicability and Feasibility

While STPA was originally developed to identify safety hazards, its applications have expanded to identification of other types of hazards, such as those related to cybersecurity (Young and Leveson, 2013; Young and Leveson, 2014) and mission success (Yoo, 2021; Scarinci et al., 2019). These identified hazards can then inform the development of system design requirements. In this chapter, we discuss the applicability and feasibility of STPA for developing technical requirements from a capability document (e.g., CDD, CPD) to accompany an RFP. To explore this, we reviewed relevant literature, held discussions with STPA users inside and outside the DAF, and applied STPA ourselves to a simple example (see Appendix C for further details). Findings from this research were augmented by discussions with systems engineers and various DAF stakeholders involved in developing technical requirements, as well as our case studies, to understand best practices for technical requirements development and DAF institutional roles, responsibilities, and constraints. See Appendix A for further methodological information and Appendix C for further details on our STPA research findings.

A Primer on Systems-Theoretic Process Analysis

STPA is a technique to identify safety issues in a system's design that could lead to unacceptable losses. It was developed, in part, to overcome the fact that many of today's systems are so complicated that past methods for evaluating their safety are inadequate (Leveson, 2012). Complex systems, such as the Space Shuttle or nuclear reactors, have many thousands of elements that interact with one another, with their human operators, and with their environments. Two existing system safety techniques, failure modes effects and criticality analysis (FMECA) and fault-tree analysis (FTA), which require the analyst to identify all possible failure modes, become increasingly untenable as the number of failure combinations that must be considered grows exponentially with the number of components. STPA takes the position that it is simpler, in some sense, to specify what the system *should* do—which is necessarily finite—than to specify all the things it should *not* do, as FMECA and FTA require, which may be quasi-infinite.

Methodologically, STPA involves a set of stakeholders and system experts working together to map out and analyze an abstract control model of the system. Depending on the complexity of the system in question, STPA may be performed separately for subsystems, to keep the number of systems engineers manageable. In addition, a facilitator who is an expert in the application of STPA is recommended, to allow the engineers to focus on the system design and behavior instead of on the analysis method itself (Leveson and Thomas, 2018). The time to complete the analysis can range from days (Montes, 2016) to weeks (per our discussions with DAF stakeholders).

Briefly, STPA begins with identification of outcomes that must be avoided in the event of system failure—e.g., loss of human lives, damage costing more than some amount, or mission failure. Next, the system is mapped out into subcomponents that interact with one another via control and feedback; an airplane control structure might include the pilot, air traffic control, the ailerons, airspeed sensors, and so forth. The engineers then scrutinize the control structure in a prescribed manner to determine potential scenarios under which the unacceptable losses identified earlier could occur.

Academic research generally supports the idea that STPA is useful for identifying potentially hazardous scenarios—typically helping the analyst to identify somewhat more unsafe scenarios than comparable techniques, such as FMECA or FTA, although STPA may require more labor (Abdulkhaleq and Wagner, 2015). It remains somewhat unclear how relevant the increased number of safety requirements STPA identifies are; a dramatically increased workload that yields only marginal safety improvements may be an inefficient use of resources. Because STPA simply counts safety requirements and does not assign probabilities to them, it is difficult to examine this trade-off quantitatively (Harkleroad, Vela, and Kuchar, 2013). Nevertheless, feedback from new STPA users tends to be positive, with many applauding STPA for helping them gain a better understanding of the system being analyzed (Abdulkhaleq and Wagner, 2015; Montes, 2016).

Systems-Theoretic Process Analysis Applicability

Insights and Enablers

Discussions with numerous stakeholders who have applied STPA highlighted several valuable insights and enablers that STPA may provide during technical requirements development. The most highly cited positive outcome from STPA was its ability to provide analysts with an understanding of the system. This understanding was often described as encompassing aspects of the system's context in terms of operational concepts, human operator behavior, boundaries and constraints, and mission profiles. While many reasons were provided for how STPA achieved this system understanding, most focused on the following aspects:

- Using STPA required analysts to shift from bottom-up (i.e., at the component-level) to top-down (i.e., at the systems level) thinking. They had to define and understand the goals of the system before performing later stages of the analysis.
- Defining these goals, developing a control structure, and identifying unsafe control actions and scenarios required analysts to do their homework on the system. To answer the questions STPA poses, analysts had to critically review relevant documents (e.g., capability requirements, operational concepts) and hold discussions with system stakeholders (operators, testers, pilots, maintainers, etc.). For example, one stakeholder

discussed STPA requiring them to wrestle with ambiguous capability statements, attempting to understand the intent of such statements.

- STPA provided these analysts with a structure and an objective when reviewing documents and holding discussions. The structured process of the analysis also allowed the analysts to organize system concepts to provide a better comprehensive picture of the system itself.

In terms of STPA outcomes, stakeholders discussed its ability to identify the system's boundaries—or "what the system should not do." This allowed analysts to build in safety or security by designing the system to prevent identified hazards from occurring. Some stakeholders also stated that STPA identified oversights in earlier analyses or processes. For instance, one stakeholder discussed being able to identify missing capability statements in requirements documents by using STPA.

STPA was also discussed as an enabler to communication and traceability throughout the acquisition life cycle. First, as mentioned earlier, the analysis requires system stakeholders to meet and discuss system testing, operation, and sustainment. Such communication promotes not only alignment of system understanding but also improved awareness and access of other stakeholders and their perspectives. Furthermore, one stakeholder mentioned the potential for STPA to create a common framework to structure communication about the system across DAF organizations throughout the acquisition life cycle. This concept draws on STPA's ability to be used at very early stages of the life cycle and iteratively refined as more information about the system is realized and on the traceability of hazards identified for mission-level goals.

Constraints and Uncertainties

While all the stakeholders we spoke with who had used STPA did proclaim its potential to usefully inform the development of technical requirements, most cautioned that the technique could not be used in isolation; one stated that STPA was not a "silver bullet." Our synthesis of these discussions with best practices in systems engineering revealed several additional analyses and tasks that may be necessary to effectively develop technical requirements.

First, STPA cannot replace much of the foundational systems engineering efforts necessary to develop technical system knowledge and to effectively write the technical requirements to be included in an RFP. Using the approach described in Chapter 3 as a framework for the comprehensive set of necessary efforts, we identified the following:

- Technical knowledge about the system must still be acquired through engineering experience and research. For example, system standards must still be identified, and market research must still be performed. Efforts to ensure the feasibility of design decisions and associated technical requirements must still be undertaken.
- Trade-off analyses will still need to be performed to decide which technical requirements are cost- or risk-beneficial.
- Design recommendations developed through STPA will still need to be translated into technical requirement statements appropriate for an RFP.

- Finally, efforts to integrate technical requirements across technical areas and ensure that requirements meet all characteristics of quality (e.g., verifiable, unambiguous, necessary) will still be needed.

Second, we could not find enough evidence to determine whether STPA can effectively perform some other important functions required for technical requirements development. For example, to perform trade-off analyses with technical requirements and to determine resource allocation and level of effort for developing specific requirements technical areas, some understanding of the relative utility or value associated with including each technical requirement is necessary (e.g., which technical requirements should be prioritized). In its traditional form, STPA produces a list of hazards that should be addressed in the design but does not provide an assessment of the risk or importance of these hazards, leaving such prioritization exercises to ad hoc processes (Harkleroad, Vela, and Kuchar, 2013). STPA literature (Harkleroad, Vela, and Kuchar, 2013; Leveson, 2019; de Souza et al., 2020) and some of the stakeholders we spoke with identified additional efforts that can be performed during STPA execution to inform such an assessment (e.g., evaluating resultant hazards with risk matrices, eliciting rankings from system operators). However, the application of these methods is immature, so evidence for their efficiency and effectiveness is lacking.

Also, we found little evidence to suggest STPA alone can derive a comprehensive set of technical requirements. As we mentioned earlier, STPA was developed to identify safety hazards by analyzing the ways a system can become unsafe and has recently been expanded to address other losses (Young and Leveson, 2013; Young and Leveson, 2014; Yoo, 2021; Scarinci et al., 2019). Hazards identified through STPA that could lead to loss of mission success may be able to inform technical requirements to address what a system should not do. The remaining question is whether defining what a system should not do can, by itself, also be inverted to define what a system should do. Theoretically, if an aircraft speed of less than x would result in a loss of mission success, an aircraft speed of x or greater could be inferred as a requirement. With little evidence of this inferred technique being used in practice, questions remain about whether such a method can capture all stakeholder needs; develop technical requirements that address mitigations if a hazard does occur; or derive all the types of technical requirements, such as those for performance, reliability, and maintainability.

As a Tool in the Toolbox

Technical requirements are traditionally developed using systems engineering (INCOSE, 2015). Discussions with systems engineering academics and SMEs and a review of authoritative systems engineering resources (e.g., DoN, 2004; DAU, 2020; ISO, 2015; NASA, 2020) revealed that various systems engineering techniques and tools can provide a number of the same insights and enablers attributed to STPA. For example, STPA insights about understanding the system and its context may also be achieved through the development of use cases (e.g., Cockburn, 2001). The development of use cases requires an analyst to define the scope and boundaries of the system, the system's stakeholders and how they interact with the system, and the goals of the system's operator. The outcome is a description of system behavior under various use and misuse conditions (e.g., system usage, operational environments) by different stakeholders. Many of the tasks our STPA stakeholders described that led to an understanding of the system are also necessary to develop use cases, including working with all stakeholders to agree on system goals; critically reviewing relevant capability documents and operational concepts; and using a structured, iterative process to define the scope, or boundary, of the design.

As another example, the benefits of using a structured process for stakeholder discussions about what the system should (and should not) do that STPA enables may also be accomplished using various requirements elicitation techniques (Zhang, 2007).[1] After the system functions have been defined, many requirements-elicitation techniques allow system requirements to be derived through structured elicitation exercises with stakeholders, e.g., using a protocol developed from the system's use cases. Requirements elicitation may also be informed by observational analysis (e.g., observing how pilots or maintainers interact with a comparable legacy system) and relevant documentation, in which information is extracted through a set of structured questions or objectives. Depending on the elicitation technique, the outputs may include performance, quality (e.g., maintainability, reliability), and safety and security.

As these examples suggest, systems engineering tools are available to accomplish many of the insights and enablers STPA provides for technical requirements development. One potential area of divergence we did find between existing systems engineering tools and STPA was STPA's ability to inform multiple approach elements in technical requirements development. In this way, STPA may be thought of as more of an end-to-end process than most systems engineering tools. STPA may play a supporting role in the following approach elements: define the context, generate technical requirements, and obtain feedback (specifically, red team or independent review).

However, given STPA's previously mentioned constraints and uncertainties, it is likely that other systems engineering tools will be needed to complete these tasks effectively and comprehensively. Furthermore, we identified tasks necessary to generate technical requirements in which STPA would not be applicable, such as informing trade-off decisions and requirements analysis. Similar to every other systems engineering tool used for technical requirements development, STPA's applicability will likely depend on the specific problem at hand and will need to be supplemented with other tools to effectively develop technical requirements. In this way, STPA may be applicable to technical requirements development as one tool in the larger technical requirements development toolbox.

---

[1] A particularly comprehensive review of requirements elicitation techniques is provided by Zhang, 2007.

Systems-Theoretic Process Analysis Feasibility

Requirements to Effectively Execute Systems-Theoretic Process Analysis

Our research identified four requirements for effectively executing STPA: (1) methodological expertise, (2) topical or technical expertise, (3) stakeholder involvement and coordination, and (4) labor and resources. STPA stakeholders stated that, if these four requirements were not met, the analyses performed could provide erroneous, unreliable, or incomplete results.[2]

Methodological Expertise

STPA was described to us both as a method and as a way of thinking. To properly perform the method, one needs to be steeped in systems thinking. Many of our STPA and systems engineering stakeholders discussed how systems thinking cannot be learned through training (e.g., a weeklong course), but instead through education (e.g., on-the-job experience, case-based learning). Current STPA training courses last one to two weeks, with guidance provided in a 188-page handbook (Leveson and Thomas, 2018). Every stakeholder who applied STPA, however, told us that the training was insufficient for them to properly perform STPA on their own.[3] Instead, many had an STPA expert (i.e., trained instructor or individual with a graduate-level education in STPA) work through the analyses with them in person and be available for follow-up questions during iterations of the analysis.

Given this ongoing need for methodological expertise and given the education and experience requirements for gaining such expertise, many STPA analyses involve an expert facilitator (Leveson and Thomas, 2018). Expert facilitation, however, does not necessarily preclude STPA participants from receiving the weeklong training. Indeed, several stakeholders we spoke with stated that at least a basic understanding of STPA is needed to be an effective participant.

Technical Expertise

STPA draws on relevant technical knowledge. The method structures that knowledge in such a way that hazards, a control structure, and unsafe scenarios can be developed. While some technical data can be gleaned from relevant documentation, system SMEs (e.g., mechanical, electrical, weapons, human factors) are needed to develop such data into useful information for STPA analyses (Leveson and Thomas, 2018). Developing the control structure requires understanding how components interact with each other, as well as with humans. Defining unsafe control actions requires experience with system design. Without such technical

[2] This is not unique to STPA. If other systems engineering tools were performed in a vacuum by untrained personnel, the results could also prove to be unreliable.

[3] See Snyder et al., 2022, for additional evidence of the limitations of STPA training.

knowledge, STPA results will be inaccurate or incomplete at best. That is, the results of STPA are only as good as the technical information used in the analysis.

Stakeholder Involvement and Coordination

Technical information is not the only input necessary for performing STPA. Others include the capabilities required; how the system will be used and could be misused; how it will be maintained; the operational environment; and other external constraints, such as compliance standards. In the DAF, this knowledge could be gleaned only through the involvement of stakeholders, including operators and maintainers. This stakeholder involvement is likely not isolated to a single discussion; instead, these stakeholders would need to be active participants in most steps of the analysis. Coordination and planning for stakeholder participation would be required, which are not trivial.

Stakeholders participating in STPA must also have some basic understanding of the method and, of course, be willing to participate. The former would likely require every stakeholder to receive at least some basic STPA training (e.g., a weeklong course). The latter, as we learned during our discussions, requires stakeholders to buy in to the analysis, which requires them to understand the value that STPA can provide.

Labor and Resources

Given that STPA is a SME-driven analysis, the resource and time requirements for its execution are mostly in the form of labor. Individuals participating in STPA need training and then will spend days to weeks performing the analysis. The time requirements increase with system complexity. One stakeholder we spoke with who had performed a simple STPA analysis estimated that it required 60 person-hours; stakeholders who had performed a much more complex analysis estimated that it took at least 600 person-hours. There are differing reports about whether STPA is more time intensive than traditional methods, with some reporting it takes up to four to six times longer (Montes, 2016). Such reports do not provide comparisons for STPA when used to develop technical requirements.

A few stakeholders who had applied STPA did state that the time required depends on the desired level of detail. Prioritizing the hazards and unsafe control actions that are most important to analyze could allow streamlining. However, they noted the difficulty of performing such a prioritization, given that no guidance was provided as part of the technique.

The DAF's Ability to Meet Systems-Theoretic Process Analysis Requirements

Determining the DAF's ability to meet the requirements for effectively executing STPA for developing technical requirements entails understanding the baseline process and environment under which the DAF currently operates. As discussed in Chapter 2, systems engineering activities in the DAF are constrained by the limited availability of expertise and resources. PMOs tend to use ad hoc processes to develop technical requirements, leaning heavily on technical requirements developed for similar programs, relevant industry standards, and SME judgment. Current available trainings focus heavily on translation activities, with little coverage of systems engineering tools, such as requirements elicitation or FFBD. In this environment, introducing STPA to develop technical requirements would likely not replace existing activities. Learning and performing STPA would, therefore, increase a PMO's workload, as would learning other unfamiliar systems engineering tools.

Our discussions with DAF stakeholders did reveal that the availability of systems engineering resources varies between programs, with some more-complex or high-profile programs being more fully manned. These programs likely perform more systems engineering activities and afford greater relevant expertise. Given our assessment that STPA would need to supplement other tools to effectively develop technical requirements, the method would likely increase the workload for these programs.

The primary impact of learning and using a new technique, such as STPA, would likely be increasing the workload of the personnel participating in the analysis. Most people we spoke with who are performing technical requirements development activities already found it challenging to perform their current duties. Any new technique might, therefore, either be performed hastily by skilled, knowledgeable individuals or be given to those without the topical expertise to provide accurate informational inputs to an analysis. The participation of all needed stakeholders would likely also require buy-in from the stakeholder organizations' leadership, which, in turn, might necessitate the participation of PMO leadership and relevant AFLCMC directorates.

An additional impact of STPA would be the need to create at least one new position type—STPA facilitators. These facilitators could be part of the AFLCMC or AFIT staffs that would move from program to program to lead discussions, capture inputs, and help develop STPA products (e.g., control structure). In concert with the PMO, these individuals would likely oversee an STPA effort, including ensuring participants are trained, facilitating discussions (to perform the STPA steps), and leading reviews of STPA products. Given the number of stakeholders and STPA steps, planning and performing each STPA assessment would be nontrivial. For a similar multistakeholder, SME-driven analysis, we developed a 60-plus page implementation guide that estimated that planning alone would take one to two months (Mayer et al., 2022). One DAF STPA stakeholder we spoke with referred to the STPA planning and training process as a "huge headache."

The creation of new STPA facilitator positions will likely require new funding. If time commitments for STPA participants (e.g., PMO personnel and other system stakeholders) are significant, additional funding may also be necessary to create new part-time roles. STPA, while itself time-consuming to perform, does not need to be performed continuously and could, therefore, be repeated quarterly or annually at the program level. Facilitators, moreover, do not necessarily need to be experts in the system in question itself—they simply guide engineers through the process and ensure that the proper steps are followed. A small handful of full-time personnel or a somewhat larger number of part-time staff would probably be sufficient to enable STPA to be performed to some level across many DAF programs. At lower levels, STPA may be used more frequently to analyze subsystems; depending on how often that takes place, more facilitators would need to be hired or trained. As discussed previously, DAF stakeholders noted that AFLCMC directorates are manned, on average, to approximately 60 percent of the current requirement before STPA. Therefore, any new funding to perform STPA would compete with meeting the remaining 40 percent of the current requirement. Overall, funding STPA efforts for technical requirements development would be challenging.

Conclusion

STPA is likely unable to fully replace current DAF efforts for developing technical requirements but could instead be a tool to use in concert with the system engineering toolbox. In less-complex or lower-profile programs, whose systems engineering activities are less mature, using STPA will likely be more burdensome for PMOs. Since current evidence of STPA's effectiveness for developing technical requirements does not exist, we are unable to determine whether the increased workload to perform STPA would prove to be cost-beneficial. Implementation of STPA for developing technical requirements for all new DAF programs could be challenging but maybe no more so than implementing any new technique requiring specialized facilitation. STPA could be among the systems engineering tools a PMO uses, but suggesting it as a systems engineering replacement is a risky proposition. While existing systems engineering tools can provide many of the same benefits and insight as STPA, systems engineering expertise and personnel are sparse and, consequently, less available for technical requirements development. The DAF would likely benefit greatly from generally expanding systems engineering expertise and resources.

Chapter 5. Discussion and Recommendations

The development of technical requirements is foundational to the success of weapon system acquisition. Ensuring technical requirements are feasible, testable, and affordable while fulfilling the documented capability gap(s) improves the probability the delivered system is suitable for operational use, while meeting schedule and budget constraints. In this chapter, we summarize our findings and a new approach for the DAF's technical requirements development and offer a set of recommendations to address the challenges we uncovered in the DAF's current process.

What We Found

Systems engineering activities to support the DAF's technical requirements development are constrained by the availability of expertise, manpower, training, and guidance. The limited organic systems engineering expertise available across the DAF is prioritized for complex and high-profile programs and allocated sparingly during the initial startup of a PMO, when technical requirements are developed. Guidance on processes and techniques for technical requirements development are not provided in DAF policy and instructions, which further lack recommended roles and responsibilities for related activities. As a result of the lack of guidance and policy and the limited systems engineering expertise and manpower at program startup, PMO engineers rely heavily on the previous experience of available personnel to develop technical requirements, technical requirements developed for similar programs, industry standards, market research, and SME judgment. PMOs' use of systems engineering tools appears to be limited, and engagement with stakeholders (operator, maintainer, etc.) is dependent on existing relationships and personalities. While the PMOs for both of our case studies engaged with the operational community, the lack of a defined role for stakeholders resulted in there being no systematic method for operational insights to be provided to, or valued by, the PMOs. Furthermore, maintenance and test community engagement was extremely limited or nonexistent. While flexibility is, indeed, important for programs as they develop technical requirements, the case studies suggest that an approach grounded in systems engineering with structured and defined opportunities for stakeholder engagement might have improved technical requirements development. That, in turn, might have prevented or mitigated some of the technical requirements oversights discovered after development contracts were awarded, such as failure to address proper positioning of the gunner's seat or pilot access to an ingress-egress ladder.

Our Approach to Technical Requirements Development

We developed a tailorable approach for technical requirements development, informed by systems engineering and the current DAF process, to overcome the various challenges discussed in the previous section. As we outlined in Chapter 3,[1] our proposed approach has seven elements (see Table 5.1). These elements and their associated tasks and stakeholders work together in a systematic, repeatable, and iterative process. Appendix D describes example systems engineering tools that can be used to perform the tasks. The approach provides a way to operationalize a set of cross-cutting themes that our research revealed.

Table 5.1. Elements of the New Approach to Technical Requirements Development

Element	Task	Stakeholders
Define context	Identify and understand stakeholders' needs Develop or understand the operational diagram	PMO Operator Maintainer Legacy system operators
Gather information	Identify system constraints and considerations Identify relevant system information	PMO Operator AFLCMC/EZ Tester Other PMOs
Develop strategy	Develop a technical requirements generation process Organize and prioritize technical areas	PMO Operator
Generate technical requirements	Define functional boundaries (expected system behavior and its boundary) Define functions (what system must do) and decompose as necessary Derive technical requirements (e.g., performance, availability, safety requirements) using appropriate derivation methods Refine technical requirements to incorporate constraints (e.g., DoD specifications) Analyze, trade off, and refine requirements (e.g., for affordability, feasibility, allocation)	PMO Operator Maintainer Tester AFLCMC/EZ
Quality check	Integrate technical requirements (identify gaps, conflicts, etc.) Translate technical requirements into contract appropriate language) Assess quality of requirements (for language, completeness, traceability etc.)	PMO Operator
Obtain feedback	Elicit and adjudicate stakeholder feedback (formal mechanism) Convene red team or independent review Elicit industry feedback (e.g., draft RFPs)	PMO Operator Maintainer Tester AFLCMC/EZ
Document	Document the process and rationale Capture lessons	PMO Operator Maintainer Tester AFLCMC/EZ

[1] Appendix D details the approach in Chapter 3 and summarized here.

The approach elements are structured to provide continual learning about the system and iterative refinement of its technical requirements. Defining the context of the operational needs and gathering information about system considerations and constraints provides the foundation for understanding the system and mission boundaries, which the field of systems engineering considers to be a crucial early step. Developing a strategy to generate technical requirements forces the PMO and system stakeholders to agree on a plan not only for technical analysis but also for how and to what extent relevant perspectives (e.g., stakeholder engagement) will be incorporated. Generating technical requirements using a top-down approach (i.e., starting with the capability needs and iteratively refining them based on constraints and considerations that are not system or mission specific) ensures that the solution is not constrained and reduces the potential for oversights in the technical requirements. All along, the various relevant stakeholders are engaged in a structured, previously agreed-on manner, and documentation is maintained. These support the ability to effectively refine technical requirements based on adjudicated decisions and justifiable rationales. The multiobjective tradespace for technical requirements development is balanced throughout the elements to generate the technical requirements, ensure their quality, and incorporate stakeholders and industry feedback. Finally, multiple opportunities are provided to ensure that the technical requirements are feasible, affordable, and meet the capability need through such tasks as requirements analysis, quality assessment, and red teaming.

Recommendations

As described previously in this chapter and in more detail in Chapter 2, the DAF currently faces challenges when developing effective technical requirements. Below, we detail a set of recommendations to prevent or mitigate the challenges revealed during our research. These recommendations are grouped temporally from short to long term.[2] The short-term ones are those that can be adopted immediately and eventually instantiated in policy. Medium-term recommendations may require additional time and funding to implement. Long-term recommendations would require new policy to initiate and permanently change the current paradigm.

Short Term

- *Adopt the proposed approach as guidance.* The approach is provided as tailorable guidance to PMOs and system stakeholders (e.g., operators, maintainers, testers). Prioritize efforts to ensure that PMOs and stakeholders understand the process. Program managers champion the approach and use it for oversight and communication. Stakeholders reference the approach to understand and ensure the various opportunities for engagement.

---

[2] All the recommendations could cause a reprioritization or reallocation of existing resources. How this is done will indicate how important successfully developing technical requirements is to acquisition leadership.

- *Assign stakeholder representatives.* Assign representatives from the implementing command and test community to the initial cadre for new programs or the PMO for existing ones. The operator representative helps ensure the technical requirements reflect the intent of the Joint Capabilities Integration and Development System requirements document and address the documented capability gap(s). The test community representative helps ensure that the technical requirements are testable. (Typically, the test community participates in existing programs; if not, a test representative should be assigned.) Once representatives have been assigned, the chief or lead engineer of the cadre or program works to integrate them into technical requirements development activities to ensure that the process is informed by stakeholder perspectives and that trade-offs in technical requirements objectives are properly adjudicated.

Medium Term

- *Supplement expertise with trusted agents.* AFLCMC/EZ, as the systems engineering home office, supplements its organic capability with a trusted agent—a systems engineering firm firewalled from competing for program participation.[3]
- *Provide and/or require additional training for PMOs.* AFIT/LS and AFLCMC/EZ develop additional training courses, specifically focused on systems engineering and related tools. Trainings are offered and advertised to engineers on cadre or PMO stand-up. Recorded virtual trainings are made available for just-in-time consumption. Additionally, appropriate personnel take formal systems engineering refresher courses (i.e., through DAU or Air Force Institute of Technology Graduate School of Engineering and Management [AFIT/EN]) at a frequency determined by the systems engineering home office.

Long Term

- *Assign formal operator roles and engagement mechanisms*:
  - Update AFI 63-101/20-101 and DAFPAM 63-128, 2008, to define formal roles for the implementing command in technical requirements development. These roles include required participation in certain activities (e.g., defining the context, requirements analysis) and/or engagement at certain milestones during technical requirements development (e.g., after receiving industry feedback, evaluation of criteria to exit process).
  - Establish adjudication processes to align PMO and operator objectives.
- *Increase organic systems engineering expertise*:
  - Create a stand-alone systems engineering series for civilians to allow the recruitment, identification, tracking, promotion, and management of this expertise as a resource.

---

[3] While the DoN has considerable systems engineering expertise, it occasionally uses trusted agents for areas in which expertise may be lacking (e.g., Mayer et al., 2019).

- Relatedly, create a Defense Acquisition Workforce Improvement Act (Title XII of Pub. L. 101-510, 1990) certification specifically for systems engineering separate from and equivalent to that of a functional or general engineer.[4]
- Consider including on-the-job training (e.g., at organizations with strong systems engineering expertise such as DoN, NASA, or relevant federally funded research and development centers) as a requirement for certification.

- *Prioritize early systems engineering during acquisition.* Consider acquisition strategy and policy changes that prioritize systems engineering similar to efforts the DoN has undertaken. Such considerations may include developing independent technical authorities that ensure system cost-effectiveness and separating weapon system contracts for contract design and detailed design and production.[5]

Our first short-term recommendation addresses how the DAF adopts our proposed approach. We recommend initially adopting it as guidance before mandating its implementation across all new DAF programs. Although the DAF can provide guidance in the near term, implementing all aspects of the proposed approach would probably require additional training, expertise, and/or manpower; new roles and responsibilities for system stakeholders (e.g., operators, maintainers, AFLCMC/EZ); and planning for additional time to develop technical requirements.

Initially, guidance should be tailored to the available expertise, manpower, and resources. Program managers should work with their chief engineers and operational representatives to identify approach elements or tasks that are feasible in the current situation. In Chapter 3, we provided a set of underlying tenets achievable, given current DAF constraints and institutional roles and responsibilities.

Many of the recommendations, including those in the medium and long-term time frames, would bolster the DAF's ability to effectively implement all aspects of our proposed approach. For example, additional training on systems engineering tools and/or an increase in organic systems engineering expertise will likely be necessary to fully perform the tasks in the generate technical requirements approach element. Furthermore, identifying stakeholder representatives and/or defining formal stakeholder roles will ensure the necessary engagement throughout technical requirements development.

As the DAF works to adopt our approach, a next step would be to test or demonstrate relevant elements or tasks on a small set of capability requirements (e.g., from a CDD or as part

---

[4] Such workforce activities could be informed by the Systems Engineering Competency Model developed as a part of an occupational analysis undertaken by the Office of Personnel Management and DoN. See Appendix J of U.S. Office of Personnel Management, 2016, and Whitcomb, Khan, and White, 2014.

[5] See Chapter Two of Mayer et al., 2019, for a review of the DoN's system design and requirements process. Organic early systems engineering work often includes more than 300 person-years of effort completed over 18 to 36 months. Contract design of ships will often be awarded to multiple offerors, who work with the DoN, followed by a down-selection to award a prime contractor for detailed design and production. Throughout the acquisition process, technical warrant holders act as authoritative experts at the systems-command level to ensure cost-effectiveness of the weapon system.

of an engineering change proposal in a post-Milestone B program). Such exploration will allow further refinement of the approach for both feasibility and effectiveness.

Overall, the recommendations presented here will allow the DAF to improve technical requirements development efforts in the near term by increasing its systems engineering footprint and, eventually, to establish a substantial organic capability.

Conclusion

While implementing the approach for technical requirements development and associated recommendations described in this report may increase early acquisition costs and schedule, the DAF will likely benefit from significant cost and schedule reductions in later phases of acquisition.[6] Reduced design, testing, and production rework; fewer design deficiencies; and improved mission effectiveness during operations could more than compensate for the initial investment. Unsurprisingly, ensuring early planning, frequent communication, and appropriate expertise will, in the end, save time and money. Many challenges with weapon system acquisition are out of the DAF's control; improving the development of technical requirements is not.

---

[6] A number of studies have shown that a significant (positive) relationship exists between systems engineering efforts and project outcomes, such as cost and schedule (e.g., Elm, 2014; Polidore, 2010; Honour, 2004).

Appendix A. Research Methodology

As discussed in Chapter 1, we used a variety of methods to develop a systems-based approach to technical requirements development. Figure A.1 illustrates the overarching research methodology. As part of the methodology, we identified three research questions:

- How does DAF currently generate technical requirements and why? (blue)
- Could STPA be used in technical requirements development? (orange)
- What are the best practices for generating technical requirements? (green)

This appendix provides a description of the methods used to answer each of these research questions.

Figure A.1. Project Research Methodology

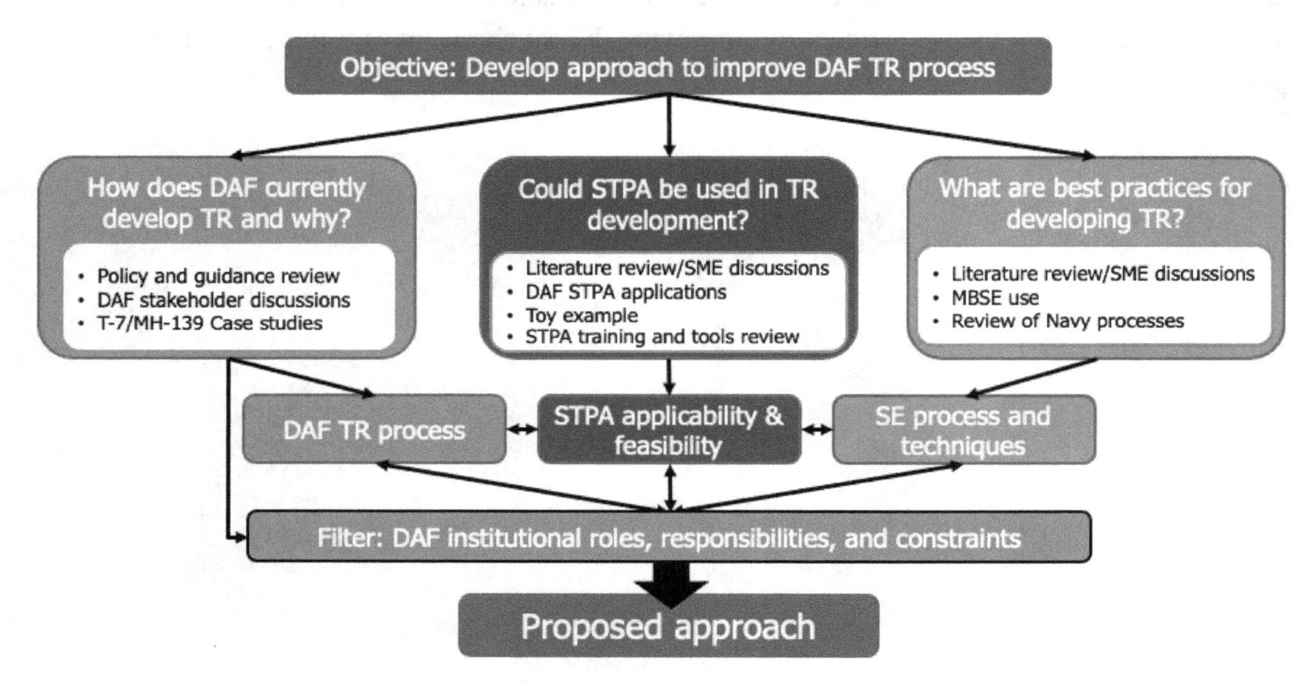

Discussions with Subject-Matter Experts

A primary form of data collection for our research involved over 40 semistructured discussions with internal DAF stakeholders and external SMEs with expertise in such areas as systems engineering, technical requirements development, and STPA. These discussions contributed data to our research and, in turn, informed the answers to our research questions. Table A.1 summarizes the discussions we held and how each informed our research questions.

Table A.1. Summary of Stakeholder and SME Discussions

Stakeholder/SME	Number of Discussions	Research Question(s) Informed
AF/A5	1	DAF technical requirements development
Academic systems engineering SMEs	5	Best practices for technical requirements development
AETC	4	DAF technical requirements development
AFGSC	2	DAF technical requirements development
AFIT/EN	1	Best practices for technical requirements development
AFIT/LS	1	Best practices for technical requirements development DAF technical requirements development
AFLCMC/EZ	2	DAF technical requirements development Best practices for technical requirements development
AFLCMC/XP	1	DAF technical requirements development
Air Force Materiel Command, Safety Office (AFMC/SES)	1	STPA use for technical requirements development DAF technical requirements development
Air Force Test Center (AFTC)	1	STPA use for technical requirements development
GAO	2	DAF technical requirements development
Ground-Based Strategic Deterrent (GBSD)	2	STPA use for technical requirements development DAF technical requirements development
MBSE stakeholders	2	Best practices for technical requirements development
MH-139 PMO	3	DAF technical requirements development
MH-139 test community	1	DAF technical requirements development
RAND systems engineering SMEs	3	Best practices for technical requirements development
SAF/AQX, SAF/AQR	2	DAF technical requirements development
STPA SMEs	5	STPA use for technical requirements development
T-7 PMO	5	DAF technical requirements development
T-7 test community	2	DAF technical requirements development
Test Pilot School (TPS)	2	STPA use for technical requirements development

Addressing How the DAF Develops Technical Requirements and Why

To answer the question of how the DAF develops technical requirements and the reasons (e.g., institutional roles, constraints) behind that, we reviewed relevant policy and guidance, conducted numerous discussions, and performed two case studies.

Policy and Guidance Review

The development of technical requirements is an important, initial step to finding vendors who can support DAF acquiring new materiel solutions. To understand the DAF's process, we sought national and DAF-level policy documents speaking to capability and requirements generation and to the acquisition life cycle. Focusing on this level allowed us to scope our review of external policies (Joint Service Specification Guides, DoDIs, military handbooks, North Atlantic Treaty Organization publications, etc.) to documents referenced by DAF policy directives and instructions. The policy review aided in understanding DAF process, roles, and responsibilities.

For the acquisition life cycle, we reviewed AFPD 63-1/20-1 and AFI 63-101/20-101 (DAF). Using these governance documents, we explored how the DAF crafts capability requirements, when technical requirements are presented to vendors, and what approaches or methods promote the transition from capability to acquisition.

Case Studies

We selected the T-7 and the MH-139 programs for case studies in concert with our sponsor. These are programs within the SAF/AQQ portfolio that were undergoing development at the time of our research. Additionally, these programs were not so far past Milestone B that their technical development process could be recounted.

Our methods involved document analysis and in-depth discussions with communities from both programs. Using semistructured discussions, we talked with PMOs, operators, and testers from both programs and reviewed their respective capability documents, SRDs, and system specifications. In the discussions, we asked respondents about their processes for developing technical requirements, any challenges they had encountered, and whether current issues could have been mitigated at the time of writing technical requirements if conditions had been different.

We used both top-down and bottom-up approaches in developing our findings. Figure A.2 illustrates these approaches. In the top-down approach, gaining an understanding of the DAF's overall process for developing technical requirements allowed us to identify techniques being used with their accompanying rationales. From this, we captured lessons learned and identified best practices to develop our findings. In the bottom-up approach, we started with the program's system specifications and traced them back to the capability requirements in the CDD or CPD. After reading the DAF program documentation,[1] we formed a basis for the past and current program status. We coupled the requirements development with the program's history and looked for linkages between the technical requirements and current issues the program may be experiencing. We selected subsystems and issues for further analysis based on document analysis (counting the number of changes through revisions) and from discussions with the PMOs and operational communities (which subsystems and issues were of concern). This top-down, bottom-up approach helped ensure that we captured as complete a picture of the program's technical requirements development as possible.

---

[1] The DAF program documentation we reviewed included the Selected Acquisition Reports, Monthly Acquisition Reports, and Program Executive Reviews (synopses of the programs' status given to the PEO by the program manager).

Figure A.2. Case Study Methodology

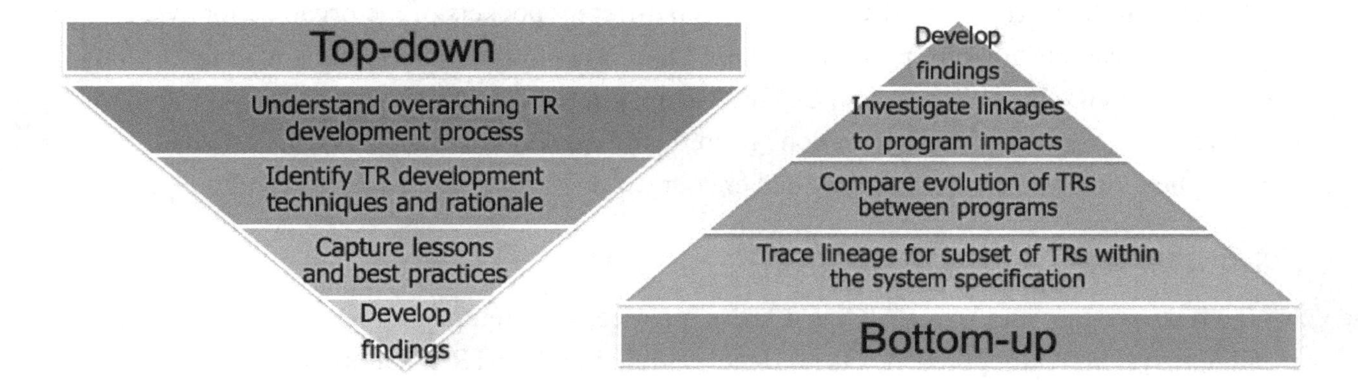

Addressing Whether Systems-Theoretic Process Analysis Can Be Used in Technical Requirements Development

To answer the question about STPA's role in technical requirements development, we conducted a series of exercises to understand STPA more thoroughly. These included a literature review, a series of discussions, a simple STPA example, and an examination of STPA training and tools.

Literature Review

We conducted a thorough literature review to further the team's understanding of STPA and, ultimately, to inform the approach presented in this report. Team members began by examining STPA at a high level, reading first through published reports on this method from its founder, Nancy Leveson. After examining Leveson's works, the review shifted to focus on work by Leveson's students. This portion of the literature review revealed three aspects of STPA that warranted further research: approach, execution, and analysis. From here, the review broadened to encompass related reports that referenced either Leveson's or her students' writing. The new reports often discussed STPA in relation to previous accidents, showing in retrospect how the results of an STPA analysis might have benefitted those affected. Some reports pointed out the technique's strengths and weaknesses, while others went as far as to offer new versions of STPA, bolstering this technique with pieces from other methods. As a part of this literature review, we exerted considerable effort to find literature discussing STPA as an application to develop technical requirements. This effort included internet searches on keywords, a snowball literature review (i.e., reviewing works cited in and references to existing papers) and requesting relevant references during all discussions with STPA stakeholders. In the end, our search revealed less than a handful of sources, with only a couple providing enough detail for analysis of STPA's applicability and feasibility for technical requirements development.

Department of the Air Force Applications of STPA

Given that the DAF is already using STPA in different pockets of the organization, we explored the insights and observations of these DAF users through discussions with three DAF organizations: GBSD, TPS, and AFTC. These semistructured discussions covered the organizations' use of STPA, observations from their experiences, outcomes and insights from their applications, and challenges they had encountered.

Toy Example

To better understand how STPA is done in practice, we attempted to perform STPA for a small, toy system. Our approach followed that outlined in the STPA Handbook, which was distributed in advance to each of the team members. After an initial meeting to identify the system, a subset of our team worked to independently perform each of the handbook's four main STPA steps, documenting our results and meeting after each step to discuss results as a group. During our meetings, we examined the notes, diagrams, and other artifacts of the STPA steps and discussed any difficulties or insights that we found during the process. We then aligned the results of each step so that the subsequent step would be based on the same inputs. After completing all steps, we held a high-level discussion to reflect on the approach, execution, and outcomes.

Systems-Theoretic Process Analysis Training and Tools

We explored various STPA software solutions listed on the MIT Partnership for Systems Approaches to Safety and Security (PSASS) website (PSASS, undated) to determine the relevance for our research and effectiveness for executing STPA. We first attempted to acquire each of the software tools listed by the MIT PSASS. As noted in Appendix C, some of these tools were not available to the public, and repeated attempts to inquire about them went unanswered. The tools we successfully acquired and installed were tested by several team members by following along with the documentation associated with each program. Because this was an exploratory analysis, we did not evaluate the ease of use and utility of each program over the course of a full STPA process; instead, we simply sought to understand the analytic features of each program.

Our understanding of STPA training requirements came from interviews with both STPA developers and practitioners and an examination of an STPA training service providers (i.e., offering lectures and facilitation of STPA) and STPA training materials (i.e., those DAF specifically uses for STPA training).[2]

---

[2] Given that STPA was developed by a relatively small number of people, mostly associated with MIT, most trainings in STPA can be traced back to a handful of experts. In fact, the STPA developers themselves offer customized training courses (taking place over the course of a few days), as well as a semiannual workshop.

Evaluating Systems-Theoretic Process Analysis Along Systems Theory Dimensions

As a conceptual exercise, we evaluated STPA along 21 dimensions of systems engineering identified in previous research (Whitehead, 2014). For each dimension, we assigned a grade from 1 (worst) to 5 (best) based on whether STPA considered the dimension as part of its process. Our scores were derived based on the knowledge of STPA that we had gained from our toy example and the official STPA literature (Leveson, 2012; Leveson and Thomas, 2018). The dimensions we considered, drawn from (Whitehead, 2014), were not exhaustive of all systems theory but we deemed them comprehensive enough to provide a sufficient comparison.

We arrived at our scores by consensus between the two team members who participated: One was more versed in the STPA literature, and the other was more versed in the systems theory dimensions. For each dimension, we considered whether the STPA process explicitly considered that dimension and to what degree. In some cases, we performed keyword searches for phrases associated with each dimension (e.g., *iteration*, *prototype*) in the *STPA Handbook* (Leveson and Thomas, 2018) to check whether STPA made note of the dimension. Of course, this was not a perfect evaluation: A good systems engineer might consider some dimensions of systems theory even if STPA does not explicitly recommend their consideration. Furthermore, while these scores were informed by reasonable knowledge of the subject matters, the scale and our inputs were necessarily subjective. Another set of researchers with similar knowledge may very well provide different answers.

Determining Best Practices for Developing Technical Requirements

Literature Review

In addition to discussions with systems engineering SMEs, we conducted a literature review of the systems engineering literature to understand best practices for technical requirements development. Our initial set of references for this literature review came from three sources: (1) the systems engineering resources provided in AFI 63-101/20-101 and DAFPAM 63-128, (2) recommended resources from our systems engineering SMEs, and (3) internet searches using relevant key words. This produced more than 165 potentially relevant references.

We prioritized and filtered this list of references by focusing on those discussing the technical requirements generation process and tools. This filtering exercise was structured by a high-level review of the references to answer the following questions:

1. Does the systems engineering reference pertain to technical requirements and mention one or more of the following aspects of requirements: generation, derivation, refinement, or verification?
2. If the answer to number 1 is yes, does the reference mention a formal or named process relating to that aspect of technical requirements?
3. If the answers to numbers 1 and 2 are both yes, does the reference describe how the formal or named process may be performed?

These questions allowed the reviewers to focus on the most promising references, narrowing the initial pool of references to one containing the 31 that were most relevant to this research. The curated set of references contained the six official and unofficial DoD- or DAF-issued guidance documents mentioned in Chapter 2 and 25 additional references spanning systems engineering guidebooks for other military and government agencies; industry and international standard organizations; and textbooks, training seminars, and peer-reviewed journal articles, many of which are referenced throughout this report.

Model-Based Systems Engineering Use

Given the increasing DAF interest in digital engineering, we also sought to specifically understand the uses and challenges associated with MBSE for technical requirements development. Internal discussions with RAND digital engineering SMEs revealed that the use of digital tools in engineering extends far beyond just MBSE models: industry and academia have developed sophisticated approaches to modeling complex systems. Therefore, we set out to understand how a variety of different DAF programs used digital tools in their engineering processes and assessments of the usefulness and challenges for technical requirements development. Based on team discussions and some brief literature reviews, we formulated a semistructured set of questions to facilitate discussions of MBSE applications with a subset of defense programs currently using digital tools. Themes from these discussions were extracted and applied to our development of the proposed approach.

Review of Navy Processes

To understand how the DoN develops technical requirements, we held high-level discussions with current Naval Air Systems Command (NAVAIR) SMEs and RAND SMEs who had recently held relevant DoN positions. As a result of these discussions, we reviewed a set of targeted documents, including the Naval Systems Engineering Guide (DoN, 2004), a recent RAND review of Navy ship design resources (Mayer et al., 2019), and NAVAIR's handbook for systems engineering technical reviews (Systems Engineering Development and Implementation Center, 2015). These discussions and literature reviews informed our long-term recommendations in Chapter 5.

Appendix B. T-7 and MH-139 Case Studies

In concert with our sponsor, we selected the T-7 and the MH-139 programs, which were parts of the SAF/AQQ portfolio that were then undergoing development. Additionally, these programs were not so far past Milestone B that their technical development process could be recounted. Overall, the case studies revealed the processes each program used to develop technical requirements, challenges encountered during the processes, the current issues and successes of the programs, and the possible relationship between these issues and successes and the technical requirements development process. This appendix provides background information on each case study program and weapon system, additional details for the findings described in Chapter 2 related to each program's technical requirements process, and the potential relationship between this process and the current status of the program.[1]

T-7A Red Hawk

Background

Currently, AETC uses T-38Cs as trainers, but with the advent of fifth-generation aircraft, e.g., the F-22 Raptor and F-35 Lightning II, AETC needed a trainer with comparable aerodynamic performance to train the next generation of pilots. This, coupled with the increasing age of its T-38C trainers, prompted AETC to initiate a trainer replacement, the Advanced Pilot Training program. The outcome of this is the T-7A Red Hawk (T-7) (Figure B.1). The T-7 will address training gaps with the T-38 and will have two components—351 aircraft and 46 Ground-Based Training Systems (GBTSs) (GAO, 2021; Secretary of the Air Force Public Affairs, 2019). Boeing was awarded an FFP contract for the program in 2018.

---

[1] The acquisition outcomes linked to contract type are from interviews with stakeholder personnel, only some of which were able to be verified in available documentation. This is a limitation of this research.

Figure B.1. T-7A Red Hawk Trainer

SOURCE: Secretary of the Air Force Public Affairs, 2019.

Requirements Development Approach

The T-7 acquisition strategy was to procure a very mature aircraft design with minimal development. This would minimize costs and expedite timelines. The T-7 PMO told us the program relied on lessons learned from past programs to develop the technical requirements. These lessons, specifically those from KC-46, led the PMO to decide to omit the SRD to prevent the contractor from making requirements changes (releasing an SRD allows the contractor to develop its own system specifications). Per stakeholder conversations, AETC and the PMO collaboratively developed the CDD and the system specifications. This collaboration could occur more seamlessly because of the concurrent development of the two documents and contributed to the PMO's confidence in its technical requirements and decision to forego the SRD. In addition, because new technology was not expected to be needed and because the contract was FFP, the technology requirements would be more straightforward than those for new development. Figure B.2 shows the technical requirements development process for the T-7 and traces the evolution from CDD to system specification to RFP then finally to contract award.

Per the discussions with the PMO, the office had used multiple resources to develop the system specification. In addition to the CDD collaboratively built with AETC, the PMO used model specifications from AFLCMC/EN, examples from other programs' specifications,[2] input from SMEs, and relevant DAF and DoD documents. Throughout system specification development, the PMO released RFIs and draft RFPs to get feedback from industry. This was helpful to determine whether the PMO was developing technical requirements contractors would

[2] These programs included the KC-46 and F-16.

Figure B.2. T-7 Technical Requirements Development Process

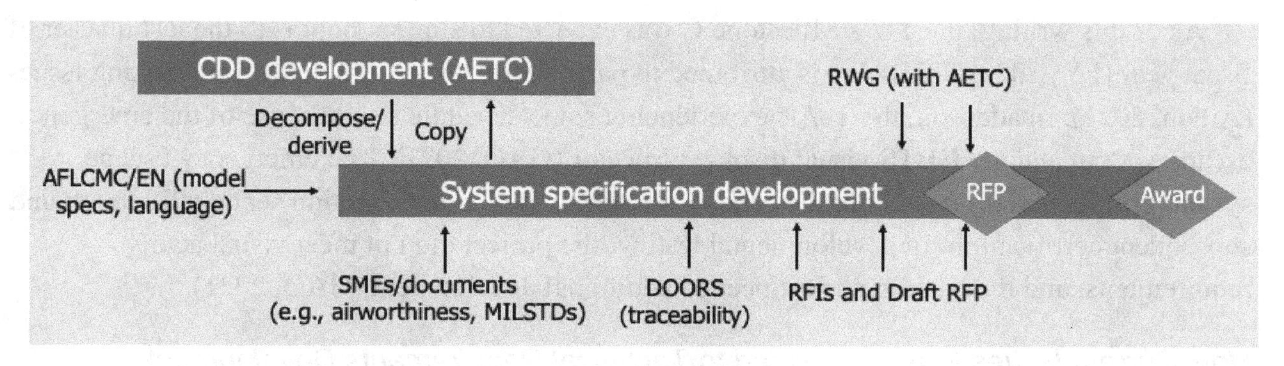

NOTE: RWG = requirements working group; MILSTD = military standard. DOORS is a requirements management tool.

not be able to meet. The PMO developed separate specifications for the aircraft (300 pages) and the GBTS (400 pages). An RWG met around six times and included AETC to ensure operator participation. The specifications were shared with industry via the draft RFPs, and the RWG adjudicated industry questions. The PMO used a company proficient in DOORS to map the requirements from the CDD to the system specification to ensure all requirements were incorporated.

Features

The T-7 system includes both the aircraft and the GBTS. The tandem-seat aircraft has twin tailfins, slats, and a shoulder-mounted wing. Its single engine generates almost three times more thrust than the dual-engine T-38C. The T-7's leading-edge root extensions aid in low-speed handling, and its retractable tricycle landing gear enables forceful braking and a straight landing path. The T-7 provides high-angle-of-attack maneuvering and high-G training. These qualities improve the handleability of the aircraft over the T-38C (Gertler, 2019; Secretary of the Air Force Public Affairs, 2019).

The T-7 training architecture integrates real aircraft (live) with ground-based (virtual) and simulated threats (constructive). This integrated live, virtual, and constructive (LVC) architecture allows aircrews to train in a more-complex threat environment. In addition, integrated LVC reduces the overall training costs because fewer aircraft are required in training scenarios. Additionally, the GBTS, a simulator system, will work with multiple systems and can mimic displays of other DAF aircraft. These simulators immerse the pilot in LVC and embedded scenarios ("Boeing Commences T-7A," 2020). GBTS has high-fidelity crew stations with dynamic motion seats and advanced high-definition projectors with 16 times more clarity than the traditional high-definition video clarity ("Boeing Commences T-7A," 2020).

Current Status

As of this writing, the T-7's Milestone C was expected to slip 15 months to the last quarter of fiscal year (FY) 2023. This delay is attributed to parts shortages, design delays, and testing issues (Albon, 2021). In addition, the T-7 is experiencing some schedule risk because of the emergency escape system and the GBTS visual display projector (GAO, 2021). The emergency escape system problems stem from both the canopy fracturing system and ejection seat qualification and subsequent certification. In developmental testing, the project did not meet visual acuity requirements, and the backup project needed additional development (GAO, 2021).

How Current Issues May Be Related to Technical Requirements Development

Some of the T-7's current issues with the emergency escape system, GBTS visual display projector, and escape and egress system (ladder) may be traced to the aircraft's technical requirements development process. The ejection seat was based on the seat qualified in the B-2 program. Per our stakeholders' discussions, the ejection seat contractor assumed modifications would be minimal to adapt to the T-7. The seat was modified, but it was also necessary to modify subsystems, reducing any benefits from using the B-2 seat. The ejection seat contractor initially estimated that ten tests (at a cost of $250,000 per test) would be sufficient for certification, but many more were required. The T-7 contract is FFP, which incentivized the prime contractor to be cost-conscious. However, more engagement with the operational and test communities might have foreseen some of the issues. AETC said the ejection seat should have been described more clearly in the system specification and wished that it had more clearly specified a modular seat design to facilitate removal and reinstallation. The normal cockpit ingress and egress was another issue and was due to the placement of the cockpit ladder. Although the requirement for an "entry and exit means that is self-contained to accommodate both cockpits" seems simplistic, it is crucial for pilots to get in and out of the aircraft safely under all conditions. The PMO learned that the contractor did not have a built-in solution. Both the PMO and the operational community thought the definition of *self-contained* was clear, but the contractor had interpreted it differently. Regarding the visual acuity requirement for the GBTS projector, the PMO knew it was pushing technology more than any previous military simulator but included the requirement at the behest of the operators.[3]

With technical requirements in general, the PMO noted that the operational community expects the replacement system to have all the features and capabilities of the legacy system and more. While communication between all the stakeholders was important, it was noted that

---

[3] The push for increased visual acuity was part of a much larger effort to "download" training from more-expensive aircraft to less-expensive simulators. While eliminating the visual acuity requirements would have streamlined the development of the T-7A, the system would no longer meet a CDD key performance parameter. It is worth noting that developing immature technology under an FFP contract is not a best practice.

deciding on the level of detail in which to write requirements was a major challenge.[4] Regular SME, systems engineering, and stakeholder engagement might have mitigated this challenge.

MH-139A Grey Wolf

The MH-139A Grey Wolf (Figure B.3) is a medium lift military helicopter that the DAF is acquiring for multiple missions, including supporting and protecting intercontinental ballistic missile (ICBM) sites and convoys, and transporting U.S. government officials and security forces in the National Capital Region, (DoD, 2018; GAO, 2021). The MH-139A will replace the UH-1N "Twin Huey," which was manufactured in 1960s. The UH-1N is being replaced because it has aging parts, can no longer meet its mission requirements, and is no longer compliant with DoD's security guidance for nuclear weapons. The MH-139 fleet of 84 aircraft is intended to fill the UH-1N's shortcomings for the AFGSC (DoD, 2018; Everstine, 2020; Tirpak, 2019).

The DAF's acquisition approach was to procure a militarized version of a commercial helicopter, Leonardo's AW139, integrated with nondevelopmental items. The Boeing Company, teamed with Leonardo, was awarded an FFP contract to develop the Grey Wolf. Although the program was considered nondevelopmental, the DAF required a system development phase to facilitate Boeing's modification of the existing helicopter, technology integration, and developmental testing.

Figure B.3. MH-139A Grey Wolf

SOURCE: King, 2020.

---

[4] AETC was told, for instance, not to provide a solution in the system specification. Therefore, such terms as *ladder*, *stairs*, and *telescoping* could not be used. AETC told us that this instruction played a critical factor in the omission of a suitable solution.

Features

Some of the features the Grey Wolf required to fulfill its missions include carrying capacity, airspeed, endurance, and survivability. The MH-139 should carry nine fully loaded troops (a total of 3,194 pounds), as required for the Minuteman III ICBM system emergency security response mission. Also, the helicopter should maintain 135 knots of true airspeed with a full load and fly for a minimum of three hours with a minimum range of 225 nautical miles without refueling. The helicopter should withstand flight critical damage for 30 minutes from a single hit, with the cockpit and cabin floor providing ballistic protection (DoD, 2018).

Requirements Development Approach

The MH-139's requirement process was different from the T-7's. As shown in Figure B.4, the MH-139 documents were created sequentially—the CPD, the RFP with SRD, and finally the system specification built by the contractor after contract award. The original acquisition strategy was to use a sole source contract to procure a COTS helicopter modified for military use. However, Congress wanted an open competition, so the PMO issued RFIs and held industry days to discuss the potential program. After industry reviewed and examined the requirements, vendors "no bid" the solicitation because they believed the requirements were not feasible. Consequently, the PMO updated the acquisition strategy including integration of nondevelopmental items with a pre-Milestone C entry ("USAF Rewriting . . . ," 2017).

Figure B.4. MH-139 Document Development Process

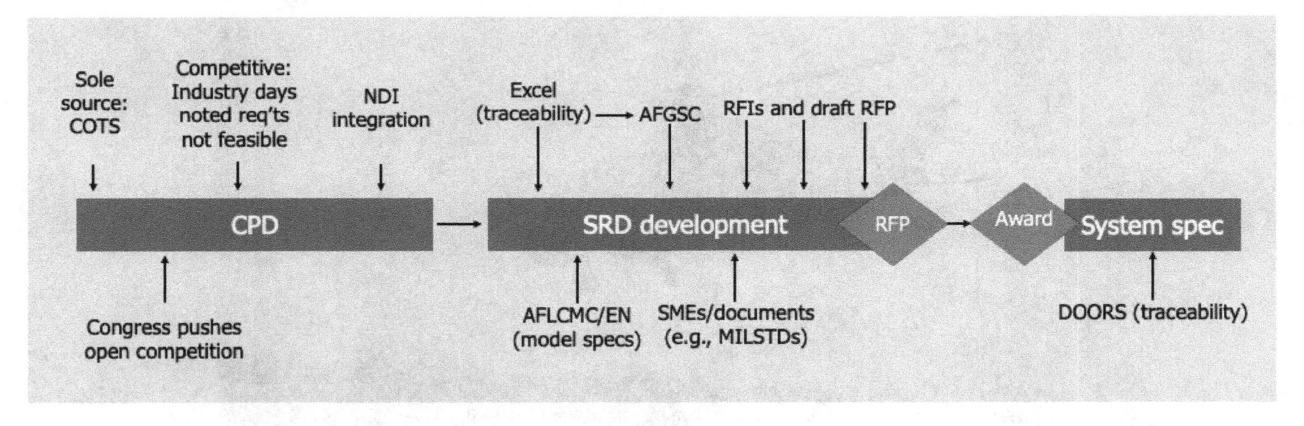

NOTES: MILSTD = military standard; NDI = nondevelopmental item. DOORS is an engineering requirements management tool from IBM.

The MH-139's technical requirements were developed to address AFGSC's identified capability gaps. AFGSC identified and refined these gaps from previous programs' requirements; nuclear security requirement compliance shortfalls; and bases, missions, and unit mission shortfalls. Once the capability gaps were documented in the CPD, the operators and acquirers collaborated in developing technical requirements. Communication between the operating community and PMO engineers involved an iterative process translating between "pilot speak" and "engineer speak" to incorporate input into the technical requirements. Sometimes the operating community sent input in pilot terms, which the engineers would translate into engineering terms and send back for review. But if the PMO was confident that it understood the operators' intent, it would not ask for feedback. Later, when issues arose, the operators asked the PMO to explain its thinking on the requirements. This inconsistent interaction contributed to the operational community's feeling its expertise had not been incorporated effectively or at all.

The initial cadre developing the program's technical requirements consisted of 12 engineers who were assigned from existing programs by AFLCMC leadership. The cadre focused on what technical requirements would satisfy the key performance parameters centered around range, payload, and endurance. The cadre's combined expertise included performance engineering, avionics integrations, systems engineering, and manufacturing.

Because the UH-1N could no longer meet its mission requirement, AFGSC had an accelerated timeline to field the new capability and, consequently, was flexible about trade-offs to shorten the schedule. Operators met with the PMO during technical requirements development with this prioritization. Per our stakeholder discussions, PMO-operator collaboration was mostly through email and telephone calls, iterating on technical requirements using Excel spreadsheets. The PMO shared the resulting technical requirements with industry, which provided feedback to the PMO, and then back to the operators. As a result of this iteration, the PMO was able to draft the SRD that accompanied the RFP.

Current Status

As of this writing, the timeline for the MH-139 to reach Milestone C was changed from the end of FY 2021 to first quarter of FY 2023. The original plan was to procure eight helicopters in FY 2021, but procurement was deferred until FY 2023. FY 2022 program funding was moved to FY 2023. This delay resulted from the Federal Aviation Administration's (FAA's) ongoing supplemental type certification issues associated with air worthiness, i.e., the weight of the helicopter and other factors. The FAA certifications are expected to be issued in FY 2022 (Reilly, 2021; GAO, 2021).

How Current Issues May Be Related to Technical Requirements Development

In addition to weight issues delaying the MH-139's FAA certification, other issues, e.g., gun and seat placement, interpretation of *austere landing capability*, and the contractor trading maneuverability for sustainability,[5] may be related to the technical requirements development process. According to our conversations, the operational community attributed its dissatisfaction

---

[5] This issue refers to the fact that the expected use of the helicopter had evolved such that it would be used more often, which increases the need for maintenance. However, the FFP logistics and parts supply contract incentivized the contractor to trade maneuverability for sustainability, exploiting underspecificity in the technical requirements.

with the gun and seat placement to how words and phrases were interpreted. How a contractor interprets technical requirements language can lead to a solution that does not meet the operators' intent. In this case, the final seat design requires the occupant to turn in the seat to shoot the gun. The technical requirement was not written incorrectly, and the solution was technically not wrong, but the human integration piece was missing. This was a problem not only with systems engineering analysis and translation but also of miscommunication. The operational community realized after the fact that asking an engineer who had never fired a gun from a helicopter to develop a technical requirement from a description of a few sentences might have contributed to the problem with the seat. A more-specific description might have improved the desired outcome.

The government also learned what constitutes austere landing capability can vary from person to person. The operator wanted to land on terrain other than a prepared landing pad, i.e., on a mountain peak, in a bog, in a cornfield, on the side of a highway, or at a missile site. The contractor did not fully understand the implications of these landings, such as a punctured tire or a chopped electrical wire. The contractor did present a study on soil depth to try to address the austere landing requirement, which the operational community did not find helpful, believing it was an engineering tool. These potential misunderstandings can arise when different communities are interpreting definitions based on their expertise rather than on what is documented.

Because the MH-139 contract is FFP, the contractor had agreed to a fixed price for logistics and parts. This potentially affects how the aircraft is operated. If it has higher usage rates than previously believed, parts may wear and degrade faster than planned, increasing the contractor's financial liability in an FFP environment. The operational community perceives that the contractor is limiting the helicopter's maneuverability, specifically banking, to lower its incurred sustainment costs. This conflict among the stakeholders (operators, acquirers, contractor) could have been adjudicated during the technical requirements development process. These concerns could have been resolved before source selection, and embedding a pilot in the PMO might have facilitated better interaction, decisionmaking, and communication. This suggests that integrating pilots with both operational and acquisition knowledge into the systems engineering process would be beneficial.

Appendix C. Systems-Theoretic Process Analysis Research

This appendix provides details about our research on STPA, which provided the foundation for much of the findings discussed in Chapter 2. For the interested reader, the appendix begins with a literature review describing STPA's approach, execution, and outcomes. We then discuss our research details, including exploration of the DAF's previous applications of STPA, a review of available STPA training and tools, insights on our execution of STPA on a toy system, and a comparison of STPA features to those of other types of systems engineering approaches.

Literature Review

STPA is an approach to system safety developed primarily by Nancy Leveson of MIT (Leveson, 2012).[1] Previous techniques—which include the likes of FTA and failure-mode and effects analysis—viewed system safety as a problem of component failure, but STPA frames the problem as one of control, or lack thereof, and has been presented as a paradigm shift in approaching system safety.

STPA is particularly guided by the fact that failures of complex systems do not necessarily require individual components to fail, but failure still results. For an illustrative but fictional example, imagine a pressurized tank holding toxic chemicals, with an emergency relief valve that vents into the open air. If the pressure in the tank gets too high, the pressure can be reduced by lowering the tank temperature. In this scenario, no staff was ever explicitly assigned the role of lowering the temperature—everyone simply assumes someone else will do it. The temperature is uncontrolled, and the pressure can build until toxic chemicals are released into the atmosphere. No individual person is responsible, and every component worked as intended, but there was clearly a design or process deficiency that should have been identified earlier. STPA aims to identify this kind of hazardous scenario by mapping out the control structure of the system in question and examining it for deficiencies.

Systems-Theoretic Process Analysis Approach

Overview

STPA is designed to be a top-down hazard-analysis technique (Leveson, 2012). The process starts by identifying losses, defined as unacceptable occurrences. These losses can then be traced through a series of steps in which the analyst sequentially determines hazards that could lead to

[1] The interested reader is encouraged to explore Dr. Leveson's textbook *Engineering a Safer World* (Leveson, 2012), as well as the *STPA Handbook* (Leveson and Thomas, 2018), which provides much more detail on the process.

each loss; unsafe control actions that may enable each hazard; and scenarios in which each unsafe control action might occur. These steps, which are useful for informing both an overarching control structure and system safety requirements, can be repeated as the system design is refined. With the control structure in hand, the analyst inspects each potential control and feedback signal to determine potential problems and develops a mitigation plan for any problems that arise. Although a single control structure for a highly complex system could be overwhelming, the STPA literature suggests it is possible to begin at a very high level and terminate the analysis once enough complexity has been reached (e.g., Leveson, 2012). Individual subsystems can then be analyzed in turn using the same process. For example, an analyst trying to apply STPA to a submarine might map out the various major components (weapon bay, nuclear reactor, sonar, and so forth) at a high level, then perform a more detailed STPA analysis on each subsystem separately.

Detailed Approach

STPA begins with the team determining the purpose of the analysis—what is the system under consideration (and what are its boundaries), and what are the unacceptable losses that the system should avoid? Examples of the latter include loss of human life, system integrity, or mission. Associated with these unacceptable losses are *hazards*, which are system states that, in a worst-case condition, will lead to an unacceptable loss. For example, if the system under consideration is a small airplane and if an unacceptable loss is the loss of human life, an associated hazard might be a person in close physical proximity to the moving airplane.

STPA's next step is to map out a control structure for the system, beginning at the highest practical level and working into more details. In this context, a control structure is simply a representation of different system components and the various commands and feedback messages that take place between them. Generating this control structure requires some knowledge of the system architecture but can plausibly be done before a design is completely finalized. STPA emphatically rejects the idea that the control structure is quantitative; instead, it is a qualitative representation that is used to guide the analyst when searching for unsafe control actions. A quantitative control diagram would require a full physics-based model of the system and would be extremely difficult to develop and/or analyze, even with modern mathematical methods.

With the control structure in hand, the analyst considers each control action in turn and analyzes four scenarios to identify whether a hazard could result: whether the signal being sent leads to a hazard; whether it not being sent leads to a hazard; whether it being sent too early, too late, or in the wrong order leads to a hazard; and whether starting it too early or stopping it too soon leads to a hazard. The results populate a table that indicates which control signals are potentially hazardous, under what circumstances they could be hazardous, and which hazards they could lead to, among other information.

The final step is to examine each of the potentially unsafe control actions discovered in the previous step and formulate loss scenarios that might occur in association with the unsafe control action. Mitigations can then be proposed that would negate the loss scenario. Consider the pressurized tank system described earlier, in which nobody is responsible for lowering the temperature of the tank. STPA might work something like this: An unacceptable loss (release of chemicals into the atmosphere) is linked to the hazard of the temperature in the tank rising above a certain value. When analyzing the control structure, the analyst notices that the control signal of "reduce temperature" could therefore be unsafe if it is not sent. When attempting to determine why the reduce-temperature command would not be sent, the analyst quickly sees that there is no reason for any individual staff member to send the command. The analyst can then recommend a mitigation, in this case, assigning a particular staff role the task of monitoring the tank pressure.

It is in this final piece of the analysis that deep system familiarity is necessary. It is not always straightforward to identify a scenario in which an unsafe control action can lead to a loss: in an example performed by the aircraft manufacturer Embraer (Scarinci et al., 2019), smoke in the cabin is traced to an air conditioner fan frets against the walls of the turbomachine, generating smoke. Without knowledge of the air conditioner design, identifying this scenario and determining a mitigation plan (presumably, realigning the fan) is difficult.

Systems-Theoretic Process Analysis Execution

Although STPA is self-contained, and a guidance handbook is available (Leveson and Thomas, 2018), our review of the literature suggests that it is not completely formalized (e.g., Harkleroad, Vela, and Kuchar, 2013). Some of the steps—particularly determining the causal scenarios that could lead to unsafe control actions—can require significant amounts of creativity on the part of the engineers (e.g., Scarinci et al., 2019, p. 17). For example, STPA does not formalize the process of generating creative scenarios; it is difficult to imagine a way any systems theory approach could. Instead, STPA helps focus engineers on where these scenarios could lead to unacceptable losses, allowing the individuals to make use of their domain expertise to envision these scenarios. This allowance for versatility raises some questions about repeatability (e.g., Asplund, El-khoury, and Törngren, 2012).[2]

In addition to STPA being unique in its approach, it also sets itself apart from other systems-theoretic techniques in its execution. STPA facilitators are recommended to properly take others though this process (Leveson and Thomas, 2018). STPA also takes somewhat more time to complete than traditional methods (Abdulkhaleq and Wagner, 2015; Montes, 2016), although the use of automation tools may improve some of these challenges.

---

[2] It should be noted, however, repeatability did not arise as a major concern during our developer and DAF application discussions.

Analysis of Systems-Theoretic Process Analysis Results

Literature has also shown that STPA derives more safety requirements than traditional methods do (Abdulkhaleq and Wagner, 2015; Montes, 2016). This, however, does not come without downsides. The technique does identify a larger number of hazardous scenarios but does not provide an adequate method to evaluate each scenario or establish priorities (Harkleroad, Vela, and Kuchar, 2013). Furthermore, the lack of a formalized STPA approach calls into question the technique's completeness. Verification of results sometimes requires using other methods, although the introduction and development of automated tools may change this in the future (Dakwat and Villani, 2018).

DAF Applications of Systems-Theoretic Process Analysis

The DAF is already using STPA in different pockets of the organization. Therefore, we decided to explore the insights and observations that DAF users offered during discussions with three DAF organizations. There was variety in organization type—a PMO, an academic unit with practical applications, and a test organization. The following sections summarize what we learned about how STPA is being employed today and present insights from organizations that have applied the technique.

Ground-Based Strategic Deterrent

GBSD is the system replacing the aging Minuteman III ICBM. GBSD consists not only of a new missile fleet but also the launch control stations, command and control systems, maintenance, cyber systems, and so forth, that are associated with ICBMs. The system is therefore immensely complex, requiring new systems engineering methodologies to help the program meet its aggressive goals for fielding the first missiles in FY 2029. The GBSD PMO turned to STPA to improve its acquisition process, hoping to save time and cost by making sure that the system that is built is the system that the warfighter actually wants—reducing the need for costly redesigns later in the life cycle.

Systems-Theoretic Process Analysis Use

With a program as large and complex as GBSD, STPA users said that it can become unwieldy to attempt to perform STPA on the entire system; instead, it could generally be applied separately to subsets of the system. Nevertheless, the users stated, a top-level GBSD STPA effort revealed a handful of simple requirements that had been overlooked in earlier drafts of the requirements (the most glaring of which was a lack of a requirement that the launch control centers actually be used to launch the missiles). Users also stated that STPA has been useful for intraprogram communication; engineers from different parts of the program who work through the process together begin to share a common lexicon and have a better sense of how the pieces of the system fit together.

Insights

Interestingly, these users did not deem the value of STPA to lie in the actual results of the process (scenario generation and requirements) but in the early stages of STPA, in which the goals of the system are examined at a high level. This forced the STPA analysts to think through the boundaries and missions of their system, which was found to be quite valuable. The other major takeaway was that STPA should be done as early as possible, to identify and correct erroneous requirements quickly.

STPA is also being viewed, in part, as a stepping-stone to a more fully realized engineering process that incorporates digital designs and advanced mathematical concepts, such as graph theory, among other elements. The use of such an integrated system, users stated, would help synchronize work across the program, reduce cost and time requirements to field large systems, and improve safety and reliability, although such a vision remains somewhat unrealized.

Test Pilot School

The DAF TPS trains students to become test pilots, flight test engineers, and flight test navigators. The 11-month curriculum, consisting of a mix of classes, lectures, and other teaching tools, leads to a master of science degree in flight test engineering. TPS has long taught a systems engineering–based approach, focusing on a predict-test-validate method that emphasizes performing tests through the lens of the scientific method.

Systems-Theoretic Process Analysis Use

STPA, as a technique, is not taught explicitly at TPS, although many of the systems theory ideas STPA builds off are taught. We learned from our discussions that the philosophy behind teaching systems theory is the increasing complexity of today's weapon systems—in particular, including cybersecurity testing meant that the testing curriculum was becoming too dense to fit into the standard 11-month program. In this context, systems theory was viewed as a flexible tool that efficiently weaves together the necessary critical thinking for students to perform all necessary tasks associated with future flight tests.

Insights

For TPS, we learned that the main benefit of STPA was its ability to force analysts to consider the bigger picture for requirements: understanding *why* a requirement exists, instead of just *what* the requirement is. This kind of critical thinking (teaching students to think for themselves, as opposed to simply training them to perform tests) was viewed as a key part of the TPS education. STPA is therefore viewed as one particular tool that can help analysts gain insights into their systems that would otherwise be missed. However, STPA has not, to date, been widely adopted in the DAF test community writ large. While it remains unclear why, we did learn that systems theory is taught, as opposed to STPA, because of STPA's resource- and time-intensive training requirements, which could not fit into the 11-month program.

Air Force Test Center

The AFTC develops and executes test plans for DAF programs. Traditionally, this has been done by leveraging the significant domain knowledge of AFTC experts. However, the existing process of building test plans is not completely systematic, relying on techniques that have worked in the past. AFTC management sought to learn whether STPA could help bring additional rigor and analysis to the development of test plans.

Systems-Theoretic Process Analysis Use

STPA has been applied thus far to develop a few test plans but is not yet widely used within AFTC. AFTC stakeholders told us that this was due, in part, to it only recently being introduced, so not enough staff have been trained in its use. Per AFTC, STPA appears to be most useful on the most complicated subset of systems, and there is not a huge demand to change the way test plans are developed. However, after some evaluations (training and tests on a few smaller systems), it is believed that STPA could be of some use in the future if momentum exists to implement it more widely.

Insights

In agreement with what other research has found, AFTC determined that STPA is potentially useful for discovering hazards and developing requirements. For many simple systems, however, traditional methods were sufficient and less time-consuming. STPA appears to be best suited for a role as an additional tool for dealing with large, complex systems. Time and manpower considerations are significant, however, and it remains unclear whether those resources could be better used in ways other than being devoted to using STPA.

Training and Tools Review

STPA, as noted previously, has been shown to require more time than other safety-analysis techniques. Evaluating a complex system also requires considering many possible combinations of component and environmental states. It is natural, therefore, to consider whether computer software can aid in performing STPA, either by reducing the time requirements or improving the clarity of the analysis or both. We explored the various software solutions MIT PSASS lists on its website (PSASS, undated) to determine whether they could improve the process of applying STPA. Our results were decidedly mixed: Most of the listed software solutions are no longer maintained by their developers or present other obstacles to their use. The software solutions we were able to use successfully aid the analyst by keeping track of different STPA artifacts, i.e., losses, hazards, and control actions. Here we briefly describe our experience evaluating each software package.

Our most successful results were with Safety Hazard Analysis Tool (SafetyHAT) (from the U.S. Department of Transportation) and the STAMP Workbench (from the Japanese government's Information-Technology Promotion Agency). Both are Windows-only programs, and while SafetyHAT appears to no longer be maintained, we were able to run and explore both. Each allows the analyst to define system components, interactions, hazards, and so forth and keeps track of which elements are linked to one another, making all results traceable to unacceptable losses. Such software is likely to be useful during an STPA analysis, to ensure the results and processes are documented and navigable, as well as being a convenient note-taking tool. Two cautions are worthy of mentioning, however. First, these types of software programs do not perform the analysis automatically—engineers are still required to evaluate the impact of different control action errors. Second, the increased level of documentation these tools generate may in fact undermine the real benefit of STPA: the process of understanding the system itself.

Computer Aided Integration of Requirements and Information Security (CAIRIS), from Bournemouth University, appeared well-documented and current, but we had difficulty with the installation process. Fortunately, a browser-based web application is available for demonstration purposes, which we explored. CAIRIS has many features similar to those of SafetyHAT and the STAMP Workbench but is aimed mainly at software developers and seems to be more fully featured than the other programs (it features the ability to easily implement different behaviors for day and night, for instance). However, the effort involved to install it may make it difficult to implement on a wide scale.

We had difficulty with the other software recommended by STAMP. eXtensible STAMP Platform (XSTAMPP), a product of the University of Stuttgart, appeared promising but presented a variety of compilation errors on installation. These apparently occurred because the code repository has not been updated in more than two years, yielding incompatibilities with more up-to-date software. The website for STPA Based Hazard and Risk Analysis Tool (SAHRA), developed by the Zurich University of Applied Sciences, appears to have been taken offline some time in 2015. Specification Tools and Requirements Methodology (SpecTRM), developed by the Safeware Corporation, is available only for demonstration on request; we received no response after multiple requests. "An STPA Tool" was built for academic research and is not available to the public. The final software package listed, Automated-STPA (A-STPA), was superseded by XSTAMPP, which as noted above, appears to be abandoned.

Other software programs for performing or aiding STPA exist outside the PSASS list, of course. We learned during our discussions with GBSD that the PMO built and uses an STPA profile in the Cameo Systems Modeler, produced by the French company No Magic, a subsidiary of Dassault. The profile is an extension to the software that allows tagging components with STPA-specific attributes and, therefore, aids in tracing requirements to unacceptable losses. Modifications to simulation software have shown promise in research as well, perhaps allowing future programs to perform some parts of the analysis automatically (de Souza et al., 2020). However, further research in this type of automation is required before it can be implemented at scale.

RAND Toy Systems-Theoretic Process Analysis Example

To become more familiar with STPA, a small RAND team worked through analyzing a toy example system. We analyzed an automatic door equipped with an occupancy sensor, of the kind one might find in a retail building. We chose this example for three reasons: first, it seemed to be a simple enough system, such that we could analyze it completely without running into issues of high dimensionality; second, it was a system familiar enough to each of the team members and no SMEs would be required; and third, it was similar to the toy example given in the systems theory training course (de Souza et al., 2020), which was also an automatic door, albeit one on a train. From the outset, we stipulated the door must open in the event of an emergency, and it should open and close when the presence or absence of a person in front of it is detected, respectively.

After settling on our example system, team members worked independently on each STPA step, convening after each step to compare results. This was done to evaluate STPA's repeatability, as well as to give each of us the opportunity to independently identify potential shortcomings. Each team member also recorded how long each step took to have a sense of the time necessary to complete the process.

We quickly found STPA demanded a more detailed description of the system than we had anticipated. In the first STPA step—identifying our unacceptable losses—we were confronted with the problem of the system boundary: Does the system include the entire building (in which case, an unacceptable loss might include theft from the building that is not prevented by the closed door), or simply the door itself (meaning unacceptable losses would be limited to the door area itself), or some middle ground? Indeed, we found ourselves asking "what is the real purpose of a door"—to *prevent* the movement of people (or air or bugs, for that matter)—or to *allow* the movement of people? In the end, we settled on three types of unacceptable losses: loss of human well-being, i.e., injury or death; loss of security, i.e., theft from the building; and loss of convenience, meaning loss of control over the airflow in the building or an inability for patrons to move through the doorway when desired.

Similar difficulties were encountered when translating our losses into hazards. Hazards, as defined by STPA, result in a loss when paired with a worst-case environmental condition. However, the distinction between hazards and losses was somewhat confusing to several team members, and subtleties of exactly what constituted a hazard were difficult to parse. Ultimately, we identified only two hazards—that the door fails to open when it should and that the door opens when it should not—which seemed to cover all our losses. For more detail, we added some subhazards as well: Failing to open when the emergency signal is received versus failing to open when a person was standing in the occupancy region were listed separately, for example.

When modeling the system's control structure, we found that different team members took different approaches: Some were fairly detailed, while others were drawn at a high level. We also encountered some confusion about the different types of control and feedback connections;

it seemed odd to draw different signals as part of the same feedback or control arrow in the diagram, given that they serve different purposes. Technical issues also cropped up, particularly with regard to the nature of the control signals. The command to open or close a door can plausibly take different forms: separate close and open signals versus a toggle signal, for example, along with the question of what the natural state of the door is: Should it stay closed except when given an explicit signal to open? Or should it remain stationary until given a signal, regardless of its current state? The exact design of the system, to our team, appeared necessary to properly determine which control actions were potentially unsafe.

On the other hand, our experience when formulating loss scenarios was generally positive, with several nontrivial scenarios (the thief pulls the fire alarm, using the emergency signal to unlock doors, for example) identified that would require thoughtful input from the customer to resolve. What remained unclear, even after performing this step, was whether every step in the STPA exercise was entirely necessary to find these nontrivial scenarios. Could the same scenarios have been found without going through the entirety of STPA? We did not resolve this question but believe it to be important to consider. Overall, the experience lent credence to the idea that STPA is better suited to evaluating the safety of a system than to designing the system from scratch.

Comparison of Systems-Theoretic Process Analysis with Systems Engineering Tools

While STPA was developed to handle the increasing complexity of today's systems, this concern is not new. Systems theory and engineering academics have long been concerned with increased complexity, and our survey of systems engineering tools (see Appendix D) found many that help engineers conceptualize different risks to their systems. This led to our team performing and internal exercise comparing STPA to systems engineering (specifically, as a collection of the various tools used to perform it). The exercise was a conceptual one and used to improve our understanding of STPA as opposed to evaluate it. Overall, we found that STPA is perhaps the most fully integrated analysis method, in that it spans several steps of the systems engineering process and works in a complementary fashion with other systems engineering tools, but that it is unlikely to stand on its own.

To perform this comparison, we evaluated STPA along 21 dimensions of systems engineering (Whitehead, 2014). For each dimension, we assigned a grade from 1 (worst) to 5 (best), based on whether STPA considered the dimension as part of its process. For example, one systems engineering dimension was *state of the system*, the idea that a systems engineer needs to analyze not just the system design or intended function but also the internal system state in different scenarios. Analyzing this dimension is strongly emphasized in STPA, where certain combinations of system states can lead to hazards and losses. We therefore issued a score of 5 for this dimension.

Across other dimensions, we found STPA more wanting. Take the *descriptive scenario* dimension: the idea that the systems engineer should consider the possible world without a built system at all. STPA deals with this dimension only in a glancing way: defining an unacceptable loss with a double negative (instead of a loss of life or component, STPA encourages defining the loss of mission, in effect). But at its heart, STPA needs to be applied to a particular system (how could one find unsafe control actions for the absence of a system?). Therefore, we gave STPA a score of 1 along this dimension.

As stated in Appendix A, this exercise, while useful for internal research efforts, was necessarily subjective, with limited handling of potential biases that may occur when performing such an analysis. Therefore, we report only the broad themes that we uncovered in this process: that STPA scores highly across many dimensions but poorly on others. Scoring poorly across certain dimensions should not be considered a limitation of STPA. While we did not perform a similar evaluation on other systems engineering tools, our expectation is many or all would fair similarly. Instead, we believe this exercise improved our understanding of where STPA is appropriate to apply in the life cycle of a system. Although useful throughout the design process, it is unlikely to stand on its own and should be used in tandem with other systems engineering tools.

Appendix D. Approach in Detail

This appendix provides a detailed description of the research team's proposed approach for developing technical requirements. While more detailed than that presented in the main report, information in this appendix is aimed to inform management and communication of technical requirements development and should provide foundational awareness for program managers and stakeholders (e.g., implementing command) about relevant objectives, tasks, and tools, allowing better oversight and stakeholder engagement.[1]

As described in Chapter 3 and shown in Figure D.1, the approach uses an iterative process to convert capability requirements (orange input arrow) to technical requirements (orange output arrow) suitable for inclusion in an RFP and consists of seven elements (blue arrows).[2] Table D.1 provides a further summary of these elements, including the tasks and stakeholders involved. The table also highlights a set of tools that can be used to perform certain tasks in the approach.[3] The set of tools presented is not meant to be comprehensive; instead, it is a cross-section to familiarize readers with the different types available. In the following sections, we first provide some general considerations for operationalizing the approach. Next, we provide further details on each element task, including the recommended roles of the different stakeholders shown in Table D.1. Then, we provide a short description for each highlighted tool, with additional references for the interested reader. Finally, given the DAF's increasing use of and interest in digital engineering, we include a short section on DAF operational considerations specifically for MBSE.

Considerations for Operationalizing the Approach

The following sections provide tailorable guidance for stakeholders identified in Table D.1 to implement the tasks using the tools also identified in the table. PMOs and technical requirements stakeholders should tailor this to their needs, resources, and the specific problem(s).

[1] We assume that an analyst implementing the technical aspects of this approach (e.g., using the systems engineering tools presented in the latter half of the appendix) would have formal engineering training and certification and therefore, have familiarity with most, if not all, of the technical tasks and tools presented in this approach.

[2] It is worth noting that while this approach is scoped for the development of technical requirements for an RFP, much of it may be useful for technical requirements development at later acquisition stages. For example, the approach may be used to develop technical requirements for additional capability needs identified through developmental or operational testing (e.g., through an engineering change proposal).

[3] We refer to these as *tools* in this report for simplicity, but they span what may be better referred to as *methods*, *tools*, and *processes*.

Figure D.1. Approach to Technical Requirements Development

Capability requirements
Define context
Gather information
Develop strategy
Iterative technical requirements development process
Generate TRs
Quality check
Obtain feedback
Exit criteria met
Technical requirements
Document

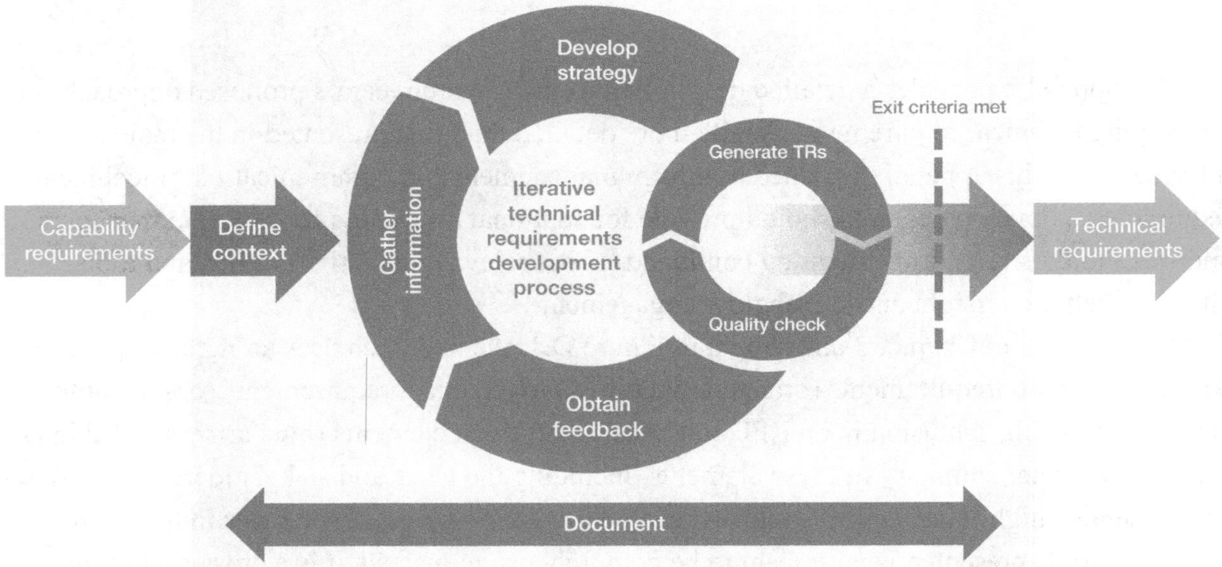

NOTES: The approach aims to convert capability requirements (orange input arrow) to technical requirements (orange output arrow) suitable for inclusion in a request for proposals. It consists of seven elements, shown as blue arrows, with each approach element being comprised of a set of tasks, tools, and stakeholders. TRs = technical requirements.

Table D.1. Approach Elements, Tasks, Stakeholders, and Tools

Element	Task	Example Tools	Stakeholders
Define context	Identify and understand stakeholders' needs Develop or understand the operational diagram	Context diagram OV-1 IDEF0 use cases STPA	PMO Operators Maintainers Legacy system operators
Gather information	Identify system constraints and considerations Identify relevant system information	Market research	PMO Operators AFLCMC/EZ Testers Other PMOs
Develop strategy	Develop a technical requirements generation process Organize and prioritize technical areas		PMO Operators
Generate technical requirements	Define functional boundaries (expected system behavior and boundaries) Define functions (what the system must do) and decompose as necessary Derive technical requirements (e.g., performance, availability, safety requirements) using appropriate derivation methods	Use cases STPA FFBD Use cases FFBD QAW STPA Bowtie diagram	PMO Operator, Maintainers Testers AFLCMC/EZ

Element	Task	Example Tools	Stakeholders
	Refine technical requirements to incorporate constraints (e.g., DoD specifications)		
	Analyze, trade off, and refine requirements (e.g., for affordability, feasibility, allocation)	Margin analysis MBSE House of quality matrix	
Quality check	Integrate technical requirements (identify gaps, conflicts, etc.)		PMO Operators
	Translate technical requirements (i.e., into contract language)	Translation rules	
	Assess quality of requirements (for language, completeness, traceability, etc.)	Quality requirements characteristics	
Obtain feedback	Solicit stakeholder feedback and adjudication (formal mechanism)		PMO Operators Maintainers, Testers AFLCMC/EZ
	Conduct red team or independent review	STPA	
	Solicit industry feedback (e.g., draft RFPs)		
Document	Document the process and rationale		PMO Operators
	Capture lessons		

NOTE: IDEF0 = Integrated Definition Method.

Certain tools, for example, are more resource-intensive than others, with some requiring special types of expertise and infrastructure. Here are some general considerations for programs as they determine which of the tools highlighted in Table D.1 to use:

- ***Stakeholder involvement.*** Effectively using most of the tools highlighted in the appendix requires some form of feedback or elicitation from system stakeholders (other than the PMO). Bowtie diagrams, FFBD, and MBSE are potentially exceptions, but even those produce outputs in which operator inputs or feedback could be useful. And while tool users can involve stakeholders informally for some tools, others require more-structured engagement, such as the QAW or STPA. In these cases, facilitated workshops may be necessary, which requires preparing workshop protocols and coordinating stakeholder participation.
- ***Tool infrastructure.*** A number of the tools we highlight use qualitative or conceptual methods (e.g., context diagram, use cases, STPA), requiring little infrastructure (e.g., software and/or hardware, data storage) other than, potentially, some means of developing a picture or diagram. Others, however, require modeling and simulation capability, computing power, or specialized software (e.g., margin analysis, MBSE).
- ***Training and expertise.*** All tools require some level of methodological training and/or expertise. Furthermore, system- and mission-specific expertise is needed to provide inputs and contextualize outputs for every tool.
- ***Tool Inputs.*** Tools differ in their input needs. Some require a basic understanding (or development) of the system design (e.g., FFBD, bowtie diagramming, STPA, margin analysis, MBSE). Some rely heavily on stakeholder or decisionmaker needs and preferences (e.g., context diagrams, use cases, house of quality matrix, market research, QAW). Others may require an initial set of technical requirements (e.g., translation, margin [trade-off] analysis, high-quality characteristics).

For programs with limited resources, manpower, expertise, or infrastructure, we recommend that any approach undertaken to develop technical requirements follow the underlying tenets described in Chapter 3 at a minimum. For convenience, the following summarizes the tenets:

- ***Develop a plan.*** While planning requires up-front time and effort, it likely saves time and resources in the long run. Define the strategy for generating technical requirements and determine how to adjudicate comments and/or disagreements from all stakeholders. Do not underestimate the level of coordination necessary to fully involve stakeholders and refine the technical requirements.
- ***Establish two-way communication.*** Opportunities for both formal and informal communication are needed between the PMO and all stakeholders. Cast a wide net for stakeholder inclusion. In the spirit of two-way communication, all stakeholders, including the PMO, should receive as much as they transmit.
- ***Adopt a continuous-learning approach.*** Initial drafts of technical requirements are a means of obtaining additional information and knowledge. Use these drafts to elicit feedback and better understand feasibility and affordability. As new knowledge is gained, updates to the strategy and technical requirements will likely be necessary.
- ***Exploit existing documentation opportunities.*** Documentation ensures that the processes used and the rationale for technical requirements decisions are captured for subsequent iterations or edifying new personnel. Existing artifacts of technical requirements development can fulfill much of the documentation needs for the process.
- ***Define the boundaries for the system and/or mission.*** Systems engineering guidance prioritizes the importance of developing a clear understanding of the system and/or mission boundaries. Answer the questions about how the system will be used; interact with its surroundings and other systems, including humans, throughout the entire life cycle; and should and should not behave in both expected and unexpected scenarios.
- ***Use a top-down approach.*** Generate technical requirements based on the capability requirements provided and iteratively refine those to address such considerations as DoD specifications, industry standards, and information from analogous systems.
- ***Balance competing needs.*** Developing technical requirements necessitates balancing many different objectives. The multiple-objective tradespace for technical requirements should be defined, and each objective should be prioritized with input for relevant stakeholders.

Implementation of the Approach Elements

This section provides a detailed description of the tasks, inputs, outputs, and stakeholder roles for each approach element.

Define Context

To define the context in which the system will operate, the PMO should identify all relevant stakeholders and understand their needs. Each stakeholder should communicate goals and priorities for the program, including how the system will be used and its interaction with the operational environment. It is very important to involve legacy or analogous system operators, maintainers, and testers in these discussions to understand potential challenges and enablers of employing the system. The understanding that results from these discussions should be captured in an operational diagram (e.g., OV-1) that details system interactions across the system's entire life cycle and discussed with the stakeholders. Additionally, the scenarios or missions under which the system will be used (i.e., use cases) should be captured.[4] Table D.2 summarizes this element.

Table D.2. Implementation of the Define Context Element

<table>
<tr><th>Task</th><th>Task Description</th><th>Inputs</th><th>Outputs</th><th>Stakeholder Roles</th></tr>
<tr><td>Identify and understand stakeholder needs</td><td>Identify all stakeholders (operator, testers, maintainers, etc.) and discuss program goals, critical capabilities, ambiguities in capability document</td><td>Capability document</td><td>Notes from stakeholder discussions
Prioritized capabilities list</td><td rowspan="2">PMO
• Identify and hold discussions with all stakeholders to understand needs and how the system will be used
• Compare stakeholder needs to the capability document and clarify any ambiguities
Operator
• Participate in PMO discussions
• Prioritize capabilities in the capability document
• Ensure representation of legacy system pilots in discussions
Maintainer
• Participate in PMO discussions</td></tr>
<tr><td>Develop or understand the operational diagram</td><td>Define how the system will be used, interactions with other systems, the environment, and/or threat systems (consider full life cycle, e.g., cradle-to grave approach)</td><td>CONOPs, CONEMPs</td><td>Operational diagram
Use case scenarios</td></tr>
</table>

NOTE: CONOP = concept of operation.

Gather Information

With the context defined, the relevant system constraints and considerations need to be identified. This requires the PMO to gather a large amount of information. AFLCMC/EZ may provide much of the information (e.g., DoD specifications) as will other stakeholders (e.g., threat systems capabilities, usage rates, maintenance philosophy), but the PMO should take the lead on market research. The analysis of the market will provide important insights into technological feasibility and affordability. The PMO should use the market research, along with other system information, to develop an integrated assessment of the system's constraints and considerations. For added context, AFLCMC/EZ should provide technical requirements from similar programs and any lessons about technical requirements development these programs documented. In addition, the PMO should hold discussions with these programs to understand whether

[4] For some programs, the tasks for defining the context may have already been performed to inform the development of the capability document (e.g., during developmental planning activities). In such cases, duplication of effort should be minimized, but the PMO and operator should ensure that personnel who were not involved in these earlier activities (e.g., some members of the PMOs' initial cadre) obtain the needed situational awareness to perform later approach tasks.

acquisition or operational impacts they are experiencing could have been prevented through technical requirements changes.[5] Table D.3 summarizes this element.

Table D.3. Implementation of the Gather Information Element

Task	Task Description	Inputs	Outputs	Stakeholder Roles
Identify system constraints and considerations	Identify possible system constraints (e.g., compliance, feasibility, and market), capabilities of interfacing and threat systems, and evolving technologies and threats	Relevant DoD specifications, industry standards, regulatory requirements, threat intelligence, details about the operational environment	• Market research analysis and conclusions • Integrated assessment of system constraints and considerations	PMO • Conduct market research to determine commercial viability and constraints on feasibility and affordability • Integrate market research findings with other inputs to develop a comprehensive list of system constraints and considerations • Hold discussions with legacy and analogous system PMO(s) to understand technical requirements development lessons AFLCMC/EZ • Provide relevant specifications, standards, and regulatory requirements • Gather and provide to PMO relevant legacy system technical requirements • Provide collected lessons from other PMOs Operator • Provide threat intelligence • Provide operational environment Tester • Provide system considerations for developmental and operational testing Legacy PMOs • Participate in discussions with PMO
Identify relevant system information	Gather technical requirements from similar programs and understand lessons from the technical requirements development processes from other relevant programs	Technical requirements from relevant programs	• Technical requirements development oversights and lessons from legacy and similar programs	

Develop Strategy

It is very important that the PMO, in collaboration with other stakeholders, develop a strategy for generating and refining the technical requirements. This strategy should account for the

---

[5] As with the previous element, some activities related to this element may have already been performed to inform the development of the capability document (e.g., during developmental planning activities). In such cases, duplication of effort should be minimized, but the PMO and operator should ensure that personnel who were not involved in these earlier activities (e.g., some members of the PMOs' initial cadre) obtain the needed situational awareness to perform later approach tasks.

unique situation the program faces, including the context, system constraints and considerations, acquisition strategy, budget, and systems engineering expertise available. While planning requires additional time at the outset, it should save time and resources in later stages of the process. The strategy should cover both technical plans (e.g., systems engineering tools and techniques) and management decisions (e.g., criteria to determine that technical requirements have met objective quality standards before being used in a final RFP). It should cover the tasks outlined in the subsequent approach elements (i.e., generate technical requirements and obtain feedback). Part of this strategy should detail how technical requirements development roles and responsibilities will be organized within the PMO and how associated resources will be allocated. The decomposition of responsibilities into technical areas will affect the system's design and, therefore, should be informed by systems engineering expertise.[6] Table D.4 summarizes this element.

Table D.4. Implementation of the Develop Strategy Element

Task	Task Description	Inputs	Outputs	Stakeholder Roles
Develop a technical requirements generation process	Develop the approach to generate and refine technical requirements (process steps, milestones, exit criteria), identify tools to manage requirements, and determine the role of stakeholders and how to incorporate feedback and adjudicate disagreements	• System context and information from previous approach elements • Program budget, acquisition strategy, expertise availability	• Technical and management strategy for generating and refining technical requirements • A clear and objective set of measures or criteria that must be met to exit the iterative development process • Planned systems engineering tools to be used that align with program constraints (e.g., budget, expertise) • Mechanisms and milestones to be used to (1) elicit and adjudicate feedback from stakeholders and (2) integrate technical requirements across technical areas	PMO • Develop technical and management strategy for generating and refining technical requirements • Elicit and incorporate stakeholder input on strategy AFLCMC/EZ • Recommend systems engineering tools to employ and provide resources (e.g., guidance, training) to support PMO use Operator, maintenance, and test • Provide inputs on strategy
Organize and prioritize technical areas	Based on stakeholder needs, available resources, and expertise, organize technical requirements development roles and responsibilities (technical areas) to align with priorities			

[6] Parts of the technical requirements strategy may have been defined prior to the initiation of technical requirements development activities (e.g., as part of the planning to stand up the PMO). In such cases, duplication of effort should be minimized, but the strategy should be revisited to ensure that it accounts for new information obtained during previous approach elements. Furthermore, the PMO and operator should ensure that personnel who were not involved in these earlier activities (e.g., some members of the PMOs' initial cadre) fully understand not only the strategy but also the rationale behind it.

Generate Technical Requirements

The technical requirements generation should align with the strategy developed in the previous approach element. While the PMO should lead all tasks in this element, frequent and early communications with the operator, maintainers, and testers is important. The PMO should employ a top-down approach to developing technical requirements; defining the functions and deriving quality attribute technical requirements based on the capability document, system context, and system and mission constraints, such as technical feasibility and affordability, environmental constraints, and the bounds of acceptable behavior of the system. Constraints and considerations that are not system- or mission-specific, such as DoD specifications, industry standards and technical requirements of analogous systems, should be incorporated after capability-informed requirements are defined. Using only specifications, standards, and requirements from analogous systems to develop the initial technical requirements for the new system increases the risk of omitting important capabilities. The derivation of technical requirements should ensure that all stakeholder needs have been considered, which often requires structured discussions among all parties. Finally, the PMO should ensure that initial technical requirements have been analyzed for affordability, feasibility, and allocation of physical qualities (e.g., power, weight). As part of this analysis, a tradespace of technical requirements to meet different cost, performance, and other goals should be created. The operational community should provide acceptable trades within the tradespace as the PMO drafts the technical requirements to be used in the next element.[7] Table D.5 summarizes this element.

Quality Check

The quality check element can be considered the first line of defense to identify potential errors or oversights in the draft technical requirements. It should follow the criteria and process outlined in the strategy for technical requirements generation and refinement. The PMO is responsible for integrating the technical requirements developed from the different technical areas and translating them into language appropriate for use in an RFP. The quality assessment could be managed by AFLCMC/EZ, with the PMO and operator providing inputs as needed. As these tasks are performed, issues with the technical requirements will be identified. Some may be easily mitigated, but others may require iterating back to the previous approach element (generate technical requirements). The PMO, with input from the operator, should determine whether the proposed mitigation affects system capability; if so, iteration should be performed. Table D.6 summarizes this element.

---

[7] This tradespace should also consider specifying requirements only to the level necessary, which aligns with the just enough, just in time concept from lean engineering. Technical requirements should be developed that use just in time information and just enough detail to communicate the desired design elements. Thus, every architecture decision that is made will constrain the design space. These decisions should only be made once enough knowledge is obtained to understand which technical requirements are essential as opposed to nice to have and the risks related to the decision. For more information on this concept, see Leffingwell, 2011.

Table D.5. Implementation of the Generate Technical Requirement Element

Task	Task Description	Inputs	Outputs	Stakeholder Roles
Define functional boundaries	• Use context and system constraints to define expected system behavior for each use case and its boundaries (in terms of system stimuli, user, environment, interactions, etc.) • Identify off-nominal system behaviors for "misuse" cases	• System context and constraints from previous approach elements	• Description or depiction of the functional boundaries	PMO • Lead all technical requirements generation activities • Elicit input and/or feedback from stakeholders • Conduct requirements analysis and present to operator for decisionmaking inputs Operator • Provide inputs for technical requirements derivation • Provide feedback on draft technical requirements • Recommend acceptable trades based on requirements analysis Maintainers and testers • Provide input for technical requirements derivation • Provide feedback on draft technical requirements • AFLCMC/EZ • Provide support and expertise, as needed • Provide feedback on draft technical requirements
Define functions	• Determine the system functions (i.e., what the system must do to meet capability needs) • Determine whether certain functions need to be decomposed and if so, decompose	• Functional boundary, capability document	• Draft functional technical requirements	
Derive technical requirements	• Derive the quality attribute technical requirements (e.g., performance, availability, safety) using appropriate methods	• Functional requirements, capability document, system context	• Draft quality attribute technical requirements • Notes from derivation activities and discussions	
Refine technical requirements incorporating constraints	• Iterate technical requirements to ensure that the system meets system constraints	• Draft technical requirements, system constraints, and considerations	• Draft technical requirements with refinements noted and explained	
Analyze, trade off, and refine technical requirements	• Analyze requirements for affordability, feasibility, allocation, etc. • Consider cost, schedule, and performance trade-offs for alternative parameter values and solutions • Refine technical requirements based on decisions	• Draft technical requirements	• Requirement analysis results • Draft technical requirements with refinements noted	

Table D.6. Implementation of the Quality Check Element

<table>
<tr><th>Task</th><th>Task Description</th><th>Inputs</th><th>Outputs</th><th>Stakeholder Roles</th></tr>
<tr><td>Integrate technical requirements</td><td>Integrate technical requirements across technical areas to identify and mitigate redundancies, conflicts, and interface gaps</td><td>Draft technical requirements from each technical area</td><td>• Identified deficiencies
• Integrated draft technical requirements with refinements noted
• Determination of whether refinements affect system capabilities</td><td rowspan="3">PMO
• Integrate and translate technical requirements
• Determine mitigations for discovered deficiencies during all tasks
• With input from operator, determine whether proposed mitigations should require iteration back to previous approach element
Operator
• Provide inputs for quality assessment
• Determine whether proposed mitigations to discovered deficiencies impact system capability
AFLCMC/EZ
• Oversee quality assessment of requirements</td></tr>
<tr><td>Translate technical requirements</td><td>Translate technical requirements into appropriate contract language</td><td>Integrated draft technical requirements</td><td>• Translated draft technical requirements with refinements noted
• Determination of whether refinements affect system capabilities</td></tr>
<tr><td>Assess quality of technical requirements</td><td>Assess individual requirements and the set of requirements to ensure that they meet quality standards (completeness, traceability, assumptions are stated, etc.)</td><td>Translated draft technical requirements</td><td>• Quality assessment results
• Draft technical requirements with refinements noted
• Determination of whether refinements affect system capabilities</td></tr>
</table>

Obtain Feedback

Once the technical requirements have met the quality standards, formal feedback should be provided to the PMO from three different viewpoints: stakeholder, independent, and industry. The process for eliciting and adjudicating feedback should follow the technical requirements development strategy. It is recommended that AFLCMC/EZ coordinate and facilitate the red team or independent review. This task can take many different forms but should include individuals already familiar with the system and mission to reduce the need for additional personnel and resources. As a part of this review, participants should question the assumptions underlying the technical requirements (e.g., feasibility, system use conditions) and review documentation on efforts to generate and analyze requirements to uncover the potential for infeasible, incorrect, inaccurate, or missing technical requirements. The three forms of feedback should result in a set of follow-up actions for the PMO and approved by the operator. As in the previous element, some follow-up actions may be easily resolved. Others, however, may necessitate additional information and development of a strategy to refine the technical requirements. The PMO, with input from the operator, should determine whether, and to what extent, further iterations are required.[8] Table D.7 summarizes this element.

Table D.7. Implementation of the Obtain Feedback Element

Task	Task Description	Inputs	Outputs	Stakeholder Roles
Elicit and adjudicate stakeholder feedback	• Formal mechanism to elicit stakeholder feedback on technical requirements • Adjudicate feedback from stakeholders as dictated by the strategy developed	Draft technical requirements	• Stakeholder feedback with adjudication decisions and follow-up actions noted • Determination of whether follow-up actions affect system capabilities	PMO • Lead stakeholder and industry feedback tasks • Review feedback from all tasks and develop follow-up actions • With input from operator, determine whether follow-up actions should require iteration back to previous elements (and which elements must be revisited) Operator • Provide stakeholder feedback • Identify personnel to participate in independent review, including a representative from the pilot community • Approve follow-up actions developed by PMO and determine whether they affect system capability AFLCMC/EZ • Oversee and coordinate independent review Test and maintenance • Provide stakeholder feedback • Identify personnel to participate in independent review
Convene red team or independent review	• Convene an independent team to review technical requirements, and identify issues and oversights	Draft technical requirements	• Independent review results • Follow-up actions based on review results • Determination of whether follow-up actions affect system capabilities	
Elicit industry feedback and adjudication	• Formal mechanism to elicit industry feedback on feasibility, affordability, potential solutions, and market interest	Draft technical requirements	• Industry responses • Adjudication decisions and follow-up actions • Determination of whether follow-up actions affect system capabilities	

Document

Documentation of the technical requirements development process and decision rationale should take place throughout. Existing artifacts can be used to reduce the resources required for this element. Outputs from each task provide much of what is needed to document the decision rationale and ensure traceability through the various iterations, decision cycles, and potential personnel changes through the course of the process. The PMO should be responsible for capturing these artifacts and storing them in a manner readily accessible to stakeholders. Only one additional form of documentation is recommended: The PMO and other stakeholders should capture lessons learned during the process. AFLCMC/EZ should collect and store this

[8] In some cases, the refinement of technical requirements may call into question the context definition developed at the start of the process. This may result in broader changes to the capability document (or its interpretation) or to the operational diagram related to the define context element. In these cases, the SPO and stakeholders may need to revisit this first task before further refining the technical requirements.

documentation on contract award for the system to be available to future programs. Table D.8 summarizes this element.

Table D.8. Implementation of the Document Element

Task	Task Description	Inputs	Outputs	Stakeholder Roles
Document the process and rationale	Throughout technical requirements development, save readily available documentation that can inform future questions about the process used and rationale for decisionmaking	Readily available documents, emails, and notes	• Organized site with documentation	PMO • Develop and organize site for documentation • Upload readily available artifacts to site • Contribute lessons Operator, maintainer, and tester • Upload readily available artifacts to site • Contribute lessons AFLCMC/EZ • Collect lessons from programs and store to provide to future programs
Capture lessons	Upon contract award for system, collect and document lessons for future programs	N/A	• Collected lessons to inform future programs' development of technical requirements	

Systems Engineering Tools

In this section, we provide an overview of each tool listed in Table D.1, organized alphabetically. For each tool, we briefly describe *what* the tool aims to do and *why*; *how* the tool works; *who* needs to be involved as the tool is deployed and used; the required *inputs* and expected *outputs* of the tool; and any *DoD considerations* for use. Throughout this description, we also cite useful references for further reading.

Bowtie Diagrams

What: *Bowtie diagrams* (Figure D.2) visualize system hazards, their causes, and ways to mitigate unwelcome outcomes by displaying the causal sequences that lead to failures and their consequences in a way that intended audiences can readily interpret (de Ruijter and Guldenmund, 2016).

Why: Understanding and assessing risks in a complex design requires thinking about the various ways that failures can occur and the ways they can be mitigated.

How: A bowtie diagram has four key parts: a critical event (sometimes called the top event or central event), which represents a key loss; a fault tree describing potential causes of the event; an event tree showing the potential outcomes of the event; and mitigations overlaid on the fault tree and event trees. The main idea of a bowtie diagram is that there are multiple ways for a critical event to occur and multiple potential consequences. Although it can summarize a significant amount of information, the bowtie diagram can be somewhat confusing to parse.

Thus, the intended audience of bowtie diagrams is analysts who are familiar with the diagram construction and can interpret it correctly.

Figure D.2. Conceptual Illustration of a Bowtie Diagram

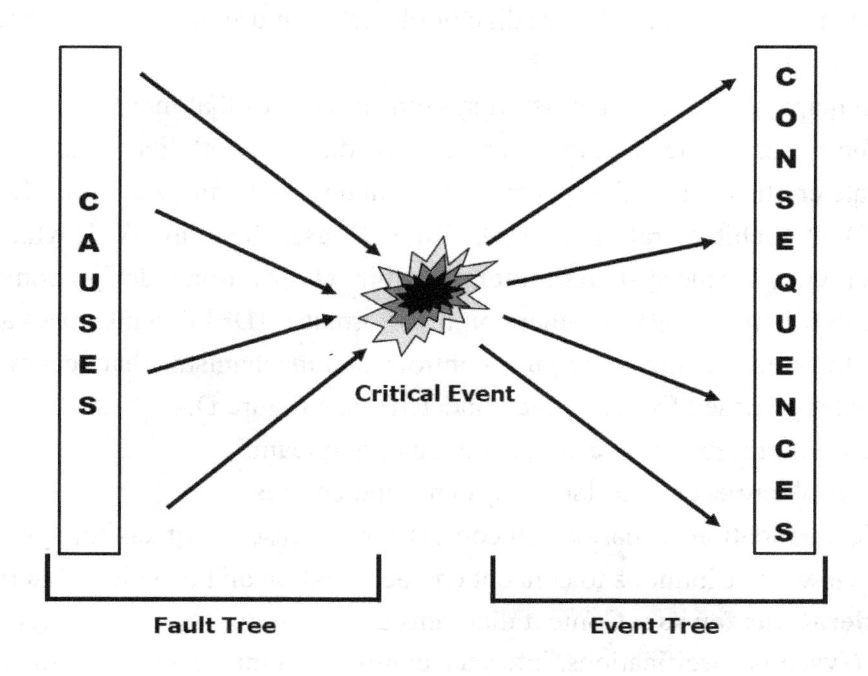

SOURCE: Adapted from de Dianous and Fiévez, 2006, p. 221.
NOTE: Not pictured are the mitigation measures that are placed on top of the fault and event trees.

Who: Technical SMEs and risk analysts familiar with bowtie diagrams.

Inputs: System design and probabilities of failure for each element in the design.

Outputs: Probabilities for failures and efficacy of mitigations imposed in reducing the failure probability. Qualitative bowtie diagrams help the analyst understand system failure modes and mitigations better but do not provide numerical outputs. However, they can still help prioritize mitigations by making failure modes more intuitive.

DoD considerations for use: Bowtie diagrams are more comprehensive than fault trees in that bowtie diagrams consider both the causes and consequences of failure. Thus, these diagrams should be used at the preliminary design stage to consider different failure mitigation methods and when performing risk audits on a design. Use of bowtie diagrams necessitates a thorough understanding of the proposed system's potential design elements, as well as how to use this systems engineering tool.

Context Diagrams, OV-1, and IDEF0

What: *Context diagrams* depict a system in a way that illustrates relationships within a specified system boundary and the external systems that may affect its functionality (NASA, 2016a; NASA, 2016b; Cyber Resiliency Office for Weapon Systems, 2020). The system's structure, behavior, requirements, and constraints can be diagrammed over a series of images. The use of context diagrams is valuable when setting up system modeling and simulation.

Why: Diagrams are broadly used as artifacts within validation activities to confirm that the intended use of a system or its constraints receive proper evaluation. The context diagram helps communicate what values stakeholders realize or obtain from use of the system and how systems engage or interact with their environment.

How: When mapping out a simple list of systems or entities that interface with the system under review, the system is the box at the center of the diagram with lines leading to external bodies as separate entities within the system's external environment (see Figure D.3 for an example). The DoD Architectural Framework (DoDAF) uses the term OV-1, which is a high-level, operational view for the system of interest within its operational design domain or priority modes of usage (SAF/AQ, 2009). Another common format is IDEF0, which uses a box-and-arrow approach to detail the inputs, outputs, controls, and mechanisms that occur at a system boundary (Knowledge Based Systems, Inc., undated). See Figure D.4.

Who: Systems engineers and the concept engineering team.

Inputs: Stakeholder needs, interfacing systems and entities.

Outputs: Test validation scenarios and context for use cases, artifacts for system requirements review, contributions to concept characterization and technical description.

DoD considerations for use: Context diagrams are common in system design documentation (systems specifications, interface control documents, etc.) and in system testing, validation, and verification protocols. In addition, the use of diagrams in any form of enterprise architectural view is a benefit helping diverse communities of stakeholders build an appreciation of competing interests when developing requirements and selecting or prioritizing viable design solutions. Stakeholder engagement when developing a context diagram is essential to ensuring

Figure D.3. A Sample Context Diagram for a Car

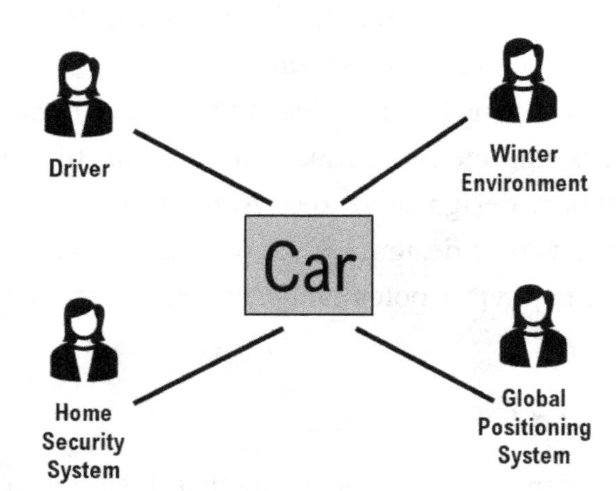

SOURCE: Adapted from Shamieh, 2011, p. 25.

proper representation of how the system interacts with other systems in the operational environment.

Figure D.4. IDEF0 Box and Arrow Graphic

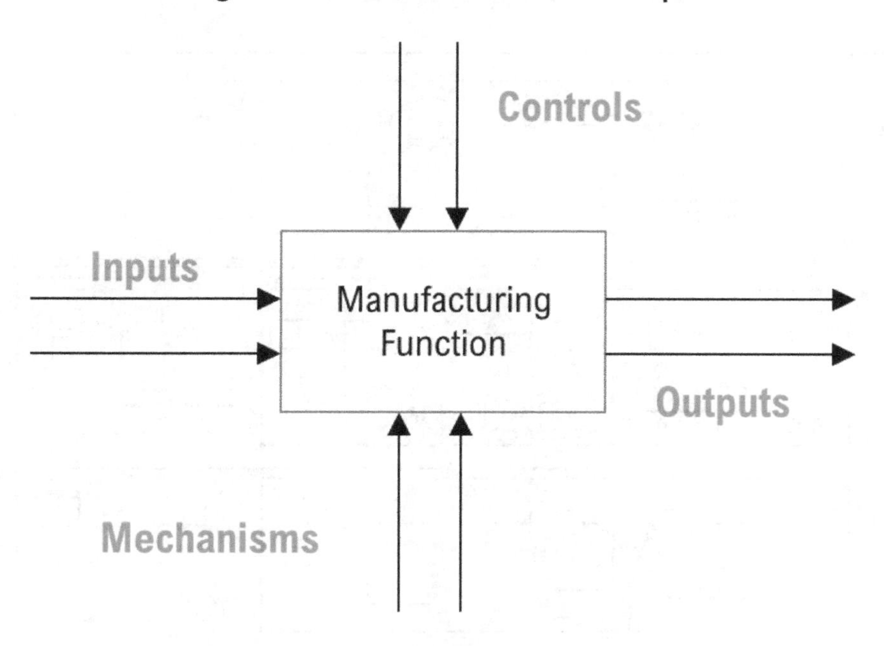

SOURCE: Adapted from Knowledge Based Systems, Inc., undated.

Functional Flow Block Diagrams

What: *FFBDs* provide relational information on the linkages, interfaces, and dependencies within a set of system functions, states, or modes of operations. As the design progresses from the abstract to specific detailed designs, FFBDs transition from the functional (e.g., series of activities within an idealized workflow) to the physical (e.g., specific named components within the overall system composition with details on the sharing of resources, transmission of energy) (Lightsey, 2001).

Why: To allow technical communities to communicate design intent with one another, other stakeholders, and end users. When accompanied with written descriptions and draft requirements statements, the diagrams provide a pictorial representation of the design details and help visualize how state capabilities or requirements interact with one another.

How: The basic elements of an FFBD include blocks, lines, and arrows. After developing a list of named blocks that represent functions, activities, or services, the designer(s) start by spatially orienting the blocks to one another. The initial layout will assume sequential activation of features or order of operation under a guidance CONOP. Lines connect the blocks with annotations when needed to describe context or requirements of the transition between functions. Finally, arrows indicate the state of the flow between each block. Many formatting features can be applied to embed extra meaning to the diagram (line thickness, dot and dash lines, color of blocks, border of blocks, etc.) (Figure D.5).

Figure D.5. Hierarchy of Functional Flow Block Diagrams

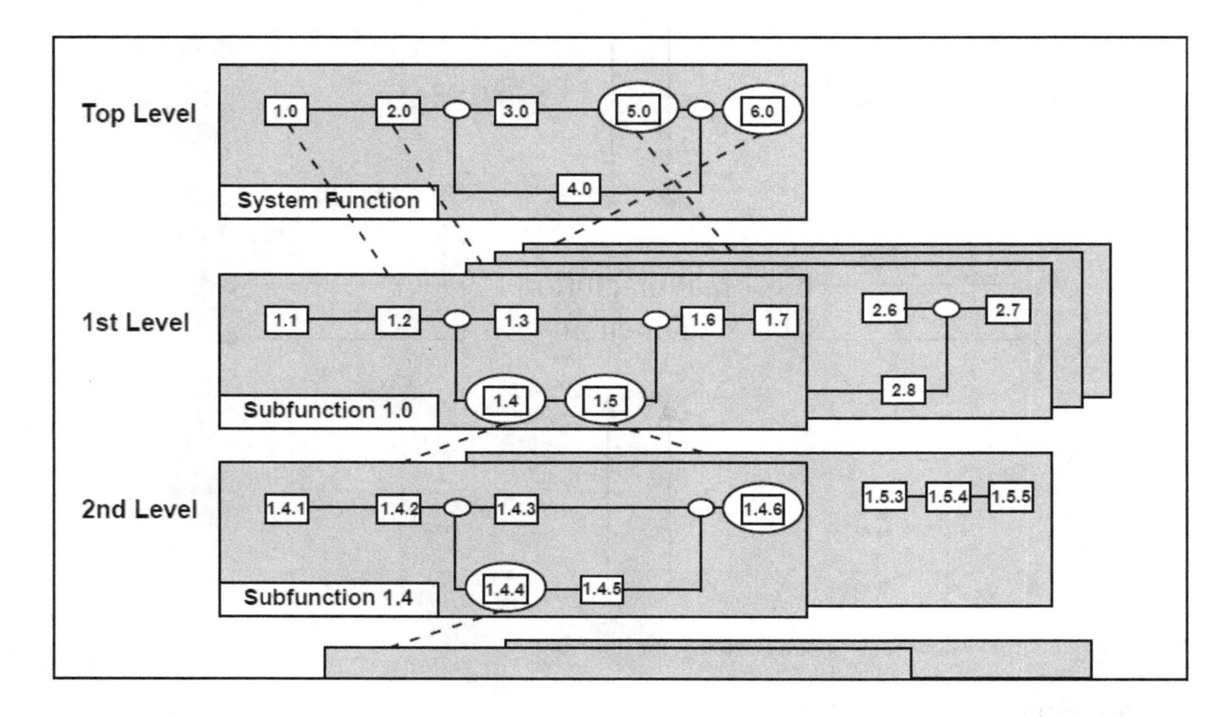

SOURCE: Adapted from Lightsey, 2001, p. 49.

Who: Reliability and maintainability engineers and contractors (DAU, 2001).

Inputs: CONOP, operational mode summary and mission profile, scenarios and vignettes, use cases, reference missions, mission threads, and DoDAF OV-1.

Outputs: Functional dependencies, estimated time sequences, and hierarchical control structures.

DoD considerations for use: FFBDs are conceptual models that detail the interfaces internal and external to the system. In abstract form, they can represent the functional intent of all design variants during an AOA. When generating design solutions, a vendor can express the final design solution as a block diagram for annotation and computational assessment of risk probabilities using empirical data from operational use (DAU, 2001). These diagrams require the engineers to have a deep understanding of the system's potential design. Developing an FFBD will also allow the designers to refine the system and maintain the necessary connections and interactions.

House of Quality Matrix

What: A *house of quality matrix*, also known as quality function deployment, is actually an aggregated set of matrices in the form of a house that relates customer needs, system attributes of a capability under consideration for acquisition and possibly competing solutions or AOAs (Figure D.6).

Figure D.6. House of Quality Matrix Example

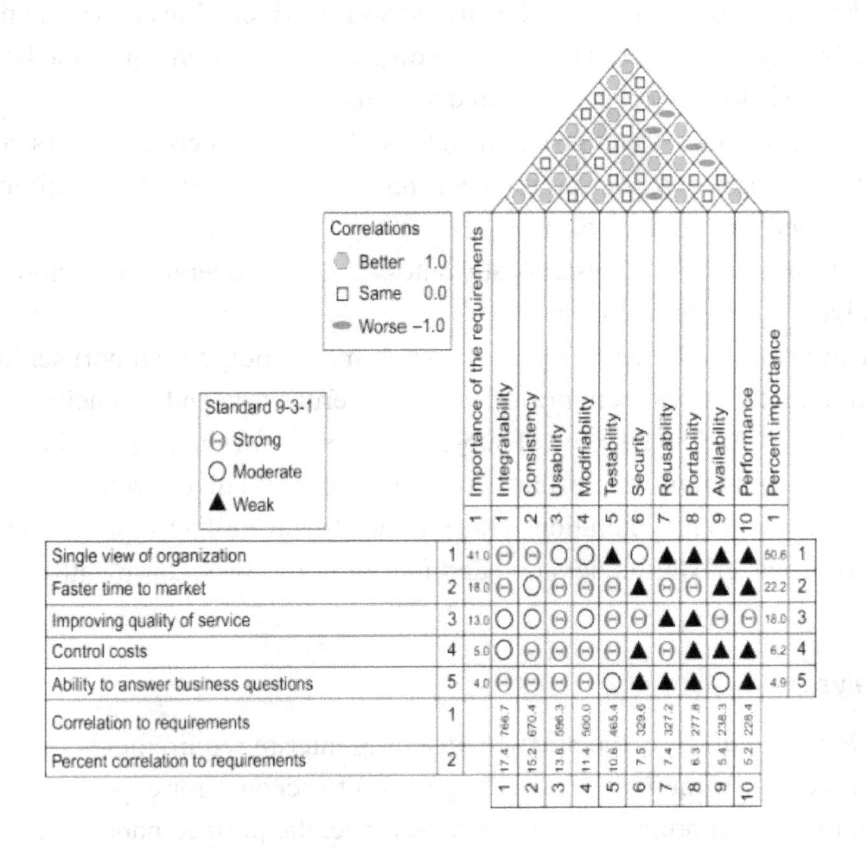

SOURCE: Image from Erder and Pureur, 2016, p. 79.

Why: The foundation of the house of quality matrix is the belief that products should be designed to reflect customer needs. The tool itself is an effective approach to connecting the voice of the customer with engineering feature selection and provides the rationale to prioritize system attributes (Nicholds et al., 2018; Wasek, Sarkani, and Mazzuchi, 2009).

How: Building a house of quality matrix involves a minimum of five steps: extracting customer needs, selecting capability attributes, identifying needs-attributes relationships, documenting trade-offs among attributes, and inputting user-based preferences.[9] Extracting customer needs involves engaging the operator community to capture the voice of the customer. These are desired experiences rather that technical performance criteria and are used to build the front door to the house. The next step is to list the mission-essential attributes and other attributes

---

[9] These steps in Figure D.6 equate to Customer Needs, System Requirements, Competition, System Specifications, and Relative Importance.

contributing to delivery of a customer need or experience, and this builds out the ceiling of the house. The living room of the house is then a relational matrix that rates the strength of association between customer needs and system attributes. In the roof of the house, engineers document the influences among system attributes as positively, negatively, or not correlated. Lastly, user-based preferences are the key to the house, which allows computation of the relative importance of system attributes. These results display in the basement of the house and allow quantitative analysis to support acquisition decisions.

Who: Systems engineers and program stakeholders, e.g., operators, users, and customers.

Inputs: Customer needs, user preferences, regulatory and statutory requirements, and competitor product performance data.

Outputs: Prioritized list of systems attributes based on coverage of customer needs and trade-off analyses for system attributes.

DoD considerations for use: House of quality matrix outputs support senior leader decisionmaking on the investment priorities for modernization and can help scope the capability document and/or SRD. Some degree of subjectivity exists for the engineering or technical community to use and draw conclusions from this tool, particularly in the early stages of capability and requirement generation. This tool needs stakeholder input and participation to ensure that priorities are well understood and faithfully followed during attribute development and trades.

Margin Analysis

What: *Margin analysis* is the practice of setting interim requirements for system qualities (mass, power, computer memory, data storage, etc.) to account for expected and unexpected growth. By allocating appropriate margins in advance, the project manager can reduce risk and increase the chance for a successful design.

Why: As system design matures, requirements for mass, electrical power, volume, or some other attribute typically grow. When writing technical requirements, this growth should be anticipated so that the final design will not exceed the physical limits of the system.

How: Margin should be calculated in a top-down manner, beginning with the overall capacity of the system, then determining the margins for each subcomponent. Typically, the overall gap between the system capacity and the current value is separated into different buckets (i.e., growth allowance, margin, reserve), and each bucket is allocated a certain amount of margin (generally 10 to 30 percent, decreasing over time as the system design matures) (NASA, 2016a; NASA, 2016b). Expertise must be applied such that margins are not overly conservative, thereby constraining the design (Naval Systems Engineering Steering Group, 2004).

Who: Program managers (to set margin policy), engineers (to compute current capacity values), and operators and maintainers (to contribute to trade-off discussions).

Inputs: Current system parameter value, expected parameter growth, and total system capacity.

Outputs: Interim bounds on parameters and acceptable growth value.

DoD considerations for use: Requirements growth is more likely in systems with lower technology readiness levels (TRLs) (NASA, 2016a; NASA, 2016b). Low TRLs are indications of technology maturity. Using immature technologies increases development and integration risk. Therefore, extra margin should be given to lower-TRL systems. As the PMO builds its acquisition strategy, it should consider TRL levels necessary to meet an operational requirement as it decides on contract type.

Market Research

What: *Market research* is a process for collecting and analyzing market information about industry's capabilities and capacity to provide technical solutions to fill the identified operational capability gap (SD-5, 2008; SD-15, 2009).

Why: This information provides the government, especially the PMO, an understanding of industry's ability to provide a technical solution off the shelf, modify an existing system, or develop a brand-new solution. This insight allows the PMO to assess the technical characteristics and feasibility of products to meet the user's needs and can inform trade-offs among system cost, schedule, and performance as technical requirements are determined.

How: Market research generally consists of seven steps, as shown in Figure D.7. While there are many ways to collect market information (e.g., business publications and web searches), RFIs allow PMOs to target interested contractors and tailor the information requests to a specific capability. Initial performance requirements can be provided to industry to test the market for feasibility and affordability. The implementing command's involvement in the development and analysis of RFI results is essential.

Who: Technical SMEs, program managers, operators, logisticians, testers, cost analysts, lawyers, and contracting officers.

Inputs: Draft technical requirements, strategic market research results (earlier RFIs), capability documents, CONOPs, and CONEMPs.

Outputs: Resulting documentation and analysis inform the government about the requirements that can be fulfilled (within cost objectives and schedule constraints) and the correct set of performance characteristics for the product.

DoD considerations for use: Market research is required by Federal Acquisition Regulations as part of the acquisition life cycle. This gives the PMO insights into potential vendors' available or future technical solutions, as well their technology's maturity. Obtaining feedback from industry on a draft RFP and associated SRD can provide the government feedback to refine its thoughts and result in improved technical requirements.

Figure D.7. The Seven Steps of Market Research

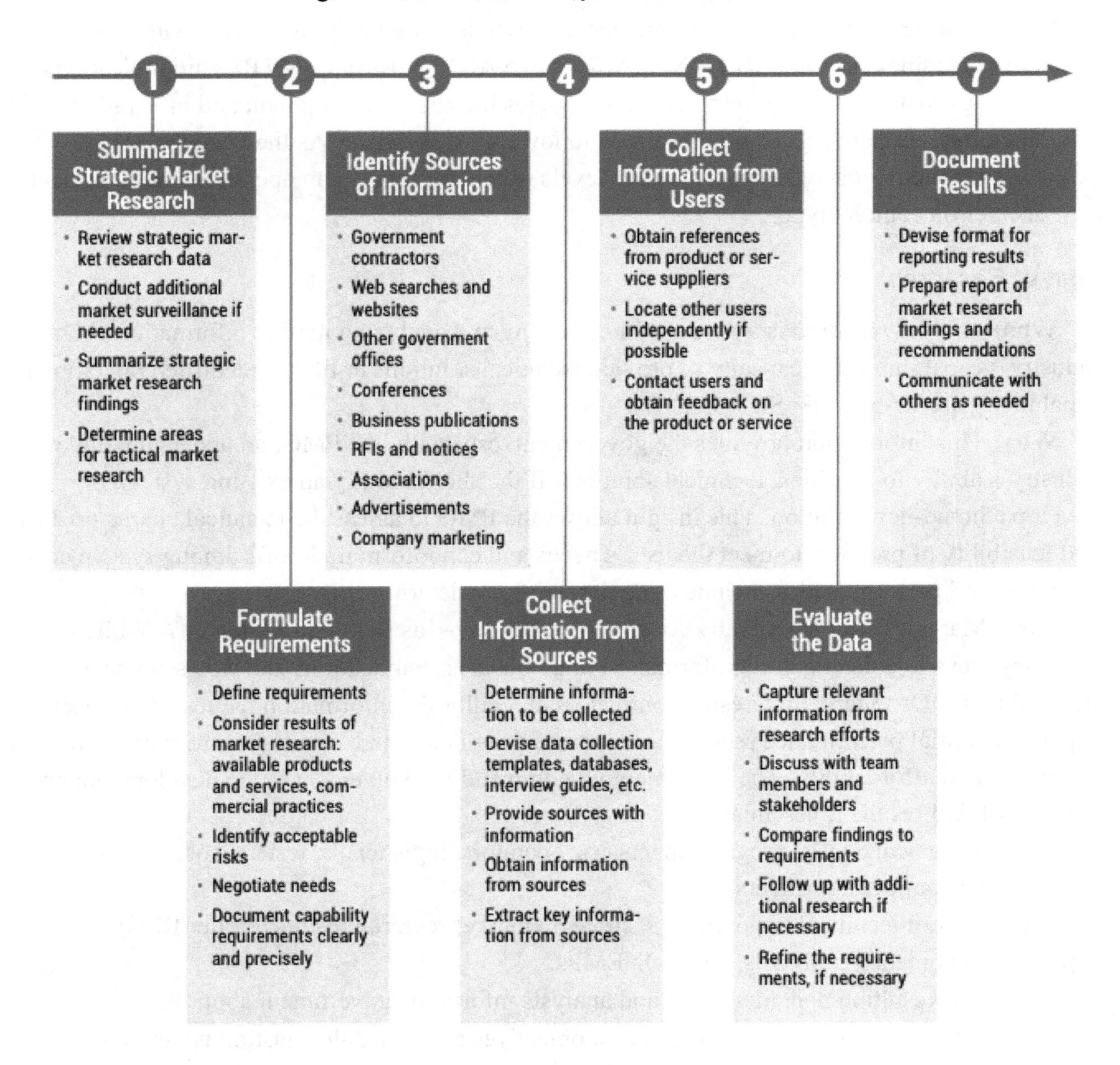

SOURCE: Adapted from SD-5, 2008, p. 14.

Model-Based Systems Engineering

What: *MBSE* is an engineering paradigm in which documents, test artifacts, and other elements are derived from a digital model (Delligatti, 2014). It stands in contrast to document-based systems engineering, in which these elements are produced individually without the benefit of a common digital model.

Why: MBSE allows faster, less-expensive iterations and updates to systems because engineering artifacts are pulled directly from the model. Updating a feature of the model once can instantly alter dozens or hundreds of documents. In addition, MBSE can reduce ambiguity in technical requirements because they can be explicitly linked to certain design elements and environmental conditions.

How: MBSE models are typically described using graphical modeling languages. Among the most commonly used are Universal Modeling Language (UML) and the UML extension, Systems Modeling Language (SysML). UML and SysML can generate several different types of engineering diagrams—structural designs, activities and sequences, use cases, and so forth. Some types of UML diagrams are functional; that is, they can be used directly to perform calculations. MBSE requires specialized software to develop and parse the UML models, along with a significant amount of training to become proficient in their use.

Who: Engineers and technical analysts who can interface directly with the models and software.

Inputs: System designs and test results.

Outputs: Derived engineering artifacts and documents.

DoD considerations for use: MBSE is an engineering paradigm that is growing in popularity and will likely become more pervasive over time. Its main benefit is to provide a centralized repository of system information, enabling faster and more-effective engineering. Because MBSE is growing in popularity, the expertise available to the DoD is somewhat limited (see the subsequent section of this appendix for more information). Additionally, there are infrastructure requirements, e.g., computing power, data storage capacity, and trained persons. DoD programs seeking to employ MBSE should include these infrastructure needs in their program planning.

Quality Attributes Workshop

What: *QAW* seeks to identify important quality attributes and clarify system requirements before an architecture has been developed (Barbacci et al., 2003). QAW is commonly used for software systems and is applicable after a software architecture has been developed.

Why: Achieving quality attributes, such as performance, security, and modifiability, is critical to achieving the success of an intended system. Requirements must therefore be developed to properly support these quality attributes and must be identified early in the system life cycle. Doing so requires an understanding of the quality attribute architectural drivers. QAW provides a way for stakeholders to analyze and agree on these drivers.

How: The method allows relevant stakeholders to gather and discuss their needs and expectations for key quality attributes, eliciting quality attribute requirements, and prioritizing quality attribute scenarios early in the system development process. Although several techniques have been designed for this purpose, they only go so far as to establish quality attribute goals. QAW recognizes that quality attribute goals alone are not definitive enough for design or evaluation and overcomes this shortfall by establishing concrete requirements.

The QAW method engages relevant stakeholders in discussion and analysis of an intended system's driving quality attributes through an eight-step workshop: QAW presentation and introductions, business and mission presentation, architectural plan presentation, identification of architectural drivers, scenario brainstorming, scenario consolidation, scenario prioritization, and scenario refinement.

The process begins with facilitators describing the process and goals of the workshop. Representatives from the stakeholder communities should then provide relevant context, as well as present the business and/or mission drivers for the system. These can be any number of high-level functional requirements, constraints, and previously established quality attribute requirements. A technical stakeholder presents a plan of the system architecture. If a detailed architecture does not exist because the QAW was held early in the system life cycle, high-level descriptions should be provided with reference to early documents. Facilitators take notes throughout these presentations, after which they generate and share a list of identified key architectural drivers. Stakeholders clarify or expand on this list before beginning to brainstorm scenarios that represent their concerns with the system. As more scenarios are shared, the group will consolidate those that are similar. This consolidation is key to prevent confusion in the next step of the process, scenario prioritization. In this step, stakeholders will be asked to vote on scenarios so that the most pressing concerns can be identified. Finally, the top four or five prioritized scenarios are further refined until participants reach agreement.

Who: Operators, maintainers, testers, trainers, architects, acquirers, systems engineers, and software engineers.

Inputs: Stakeholder feedback, business and/or mission drivers, system descriptions, context drawings, business and mission plans, and key technical requirements and constraints.

Outputs: List of architectural drivers, prioritized list of raw scenarios, refined scenarios, input for architecture design, refined requirements, case diagrams, and context diagrams.

DoD considerations for use: QAW is system-centric and stakeholder-focused. Five to 30 stakeholders should participate, and each receives a participant handbook prior to the workshop. This handbook provides examples of quality attribute taxonomies, relevant questions, and quality attribute scenarios. It is also important to note that this process is intense and demanding. All participants must stay focused and must actively participate to achieve success. Because stakeholder engagement is foundational, the government will need to give the workshop enough priority to ensure the participation of needed stakeholder.

High-Quality Requirement Characteristics

What: This tool consists of creating a list of questions or checks that can be used to ensure and establish high-quality requirements (INCOSE, 2017; MIL-HDBK-520A, 2011; MITRE Corporation, 2014; NASA, 2020). High-quality requirements are necessary, unique, clear and concise, operationally effective, feasible, traceable, measurable, complete, consistent, implementation-independent, verifiable, and singular (i.e., relate to one operational need). This list establishes a baseline for ensuring that requirements are written with high-quality.

Why: Requirements are vital in the development of intended systems and, as such, should be of high quality. Not striving for this standard can result in improperly defined requirements, leading to problems further in the system life cycle.

How: To establish that requirements are of high quality, stakeholders should review requirements with reference to this list and ensure that each requirement meets the listed criteria. Trade-offs will frequently be necessary in the development process. This does not obviate these criteria. After trade-offs have been made, it is important to assess the resulting requirements against this list to ensure that the new requirement is still high quality.

Who: Program stakeholders, SMEs, and contractors.

Inputs: System objectives, CONOPs, CONEMPs, capability gaps, stakeholder input, and draft requirements.

Outputs: Defined, unambiguous requirements.

DoD considerations for use: This technique requires systems engineers with a strong understanding of the criteria for a good requirement. Coupled with stakeholder input, requirements should be iterated until these criteria are met. Adhering to these criteria will reduce the likelihood of adverse acquisition and operational outcomes during development and operations, respectively.

Systems-Theoretic Process Analysis

What: *STPA* is a systematic method originally developed to aid in the identification of hazards in complex systems (Leveson, 2012; Scarinci et al., 2019; Leveson and Thomas, 2018). The theory behind STPA is that hazardous scenarios take place because of improper control of the system; it therefore recommends mitigating hazards by introducing controls or changing the way the system is controlled. In going through the process, the STPA practitioner is asked to determine the unacceptable losses to mitigate, then map out a high-level diagram of the system. This is then analyzed to determine ways different control actions could lead to a loss.

Why: STPA was developed, in part, to address the shortcomings of other methods of hazard analysis that looked only for hazards associated with individual component failures and, therefore, miss more subtle system-level hazards. By contrast, STPA emphasizes studying the interactions between components and human operators as a possible source of system failures. It also has the benefit of being highly traceable, allowing the reasoning behind all mitigation measures to be easily understood.

How: In practice, STPA is performed by gathering the relevant stakeholders together to work through the known system design. The process is fairly well-documented, although it is highly recommended that a trained facilitator be present to ensure that the steps are carried out as specified. In practice, STPA takes time to go through—often on the order of weeks to deal with a system of low to moderate complexity.

Who: STPA facilitator, stakeholders, and SMEs (particularly engineers who understand the system's behavior).

Inputs: Some level of system design is required; although the design need not be finalized, the more detailed it is, the more specific the recommendations will be.

Outputs: A series of hazardous scenarios associated with the system along with proposed mitigation measures.

DoD considerations for use: STPA applies many best practices from the systems engineering literature but differs from other tools in that it is an end-to-end process for dealing with hazards. Since a facilitator well-versed in STPA is recommended to implement and moderate the process, DoD will need to train and have available personnel to implement this tool. It appears that STPA can require more time and resources than other methods, albeit for results that are generally seen as more comprehensive.

Translation Rules

What: *Translation* is a technique used to convert one type of requirements to another (Phipps, 2021; OUSD, 2008; MIL-HDBK-520A, 2011; MITRE Corporation, 2014; INCOSE, 2017). Translation can be used to convert user requirements to performance requirements and then performance requirements to technical design requirement, although its most relevant use in this process is to translate technical requirements into contract language.

Why: Written requirements often follow a specific format based on the type of requirement. This format makes it difficult to include technical requirements, as written, in the contract. It is thus necessary to convert the technical requirements into contract language and ensure that the understanding of the system established by the requirement is not lost.

How: The starting point is a set of high-quality technical requirements (i.e., outputs from the High-Quality Requirements Characteristics tool) and a shared understanding of them. Stakeholders decide on the type of language to describe the requirements in the contract. The group reviews the technical requirements and, through discussion, determines how these requirements are best represented using the contract language.

Who: Relevant stakeholders, operators, SMEs, and contracting officers.

Inputs: Technical requirements, capability objectives, and stakeholder input.

Outputs: Contracting language and feedback for system design.

DoD considerations for use: Translation relies heavily on stakeholder discussion. Prior to use, DoD should consider identifying all relevant stakeholders and involving all relevant parties in such discussions. It is important to have the contracting officer involved in the discussions. This provides context and helps the contracting officer understand the intent of the capability gaps the technical requirements are fulfilling.

Use Cases

What: A *use case* is a description of system behavior under various use and misuse conditions (e.g., system usage, operational environments) by different stakeholders (Cockburn, 2001). The formats of use cases can vary—they can consist of documents, graphics, videos, models, and/or simulations. Figure D.8 is an example of a use case diagram.

Figure D.8. Simple Use Case Diagram

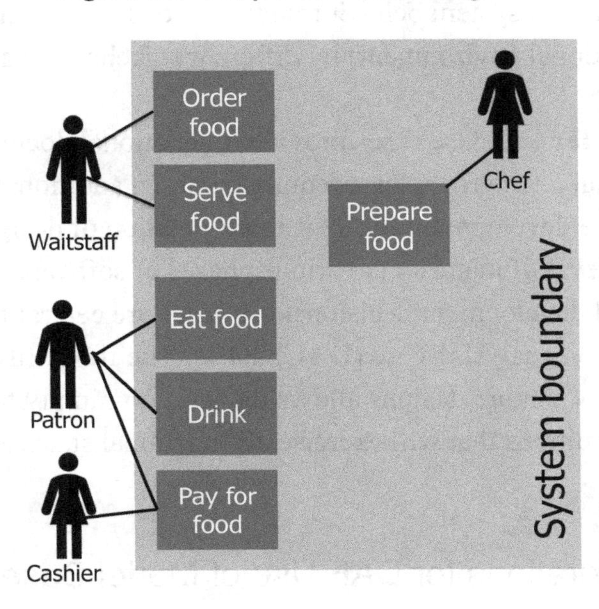

SOURCE: Adapted from Image from MITRE Corporation, 2014, p. 338.

Why: Having well-defined use cases as part of technical requirements development promotes better communication of operator needs to the solution developer. Use cases bundle user concerns within a context of system usage or operational environment. For generating system requirements, the purpose of developing a use case is to list the stakeholders and describe the value they expect to derive from using the resulting system (Shamieh, 2011). Much as a comic book tells a story or as a storyboard organizes the ideas to be presented within a movie, use cases lay out a collection of details that embody the capability needs the operating command has expressed. For example, the validation and test community relies on use cases to capture the essence of test conditions and the criteria required to verify system performance against intended requirements.

How: To develop a use case, an analyst defines the scope and boundaries of the system, the system's stakeholders and how they interact with the system, and the goals of the system's user. Some assumptions are required for crafting a use case, and some assumed dependencies can be discovered simply through elaboration of written guidance on CONOPs and draft requirements. Use cases can animate alternative outcomes from a CONOP or can remain high level, with a focus on system boundary conditions. Use cases can be designed for nominal and off-nominal conditions. With a variety of formatting standards available to design a use case, most seek to identify actors, demarcate system boundaries, enumerate key activities and events, and annotate contextual or environmental factors that influence the highlighted events (MITRE Corporation, 2014).

Who: systems engineers, with input from operators and maintainers.

Inputs: CONOPs, operational mode summary, and mission profile.

Outputs: A description of system behavior under various use and misuse conditions (e.g., system utilization, operational environment) by different stakeholders and documentation for a requirements document.

DoD considerations for use: Use cases provide foundational documentation for system- or subsystem-level functional and performance requirements. In addition, they are listed among best practices for software development for their representation of design progress toward desired capabilities and level of maturity at various phases of software deployment. Of particular importance after materiel development decisions, use cases are central to operational analysis during the materiel solution analysis phase (DAU, 2001). The flexibility of use cases to demonstrate preliminary or mature designs allows the test community to participate in early considerations of test conditions that will exercise all functional states and modes of operations of a hypothetical system.

Operational Considerations for DAF Use of Model-Based Systems Engineering

Given DAF's increasing use of and interest in digital engineering, this section includes a brief discussion of MBSE-related operational considerations for the DAF. The insights described were integrated through discussions with DoD programs that have varying levels of experience and expertise with using MBSE.

Model-Based Systems Engineering Infrastructure and Expertise

MBSE relies heavily on two components: complex software programs (such as Dassault's Cameo or IBM's Rational Rhapsody) and experienced operators of the software. Both present difficulties for DAF programs.

Software and Infrastructure

First, software licenses are required for any computer running an MBSE program, and computing environments must be properly configured to deal with different levels of classification. These can be costly investments to make, and unless the PMO is fully committed to the MBSE paradigm, these software and hardware investments can end up being underused.

Additionally, because implementing this infrastructure often involves significant lead times, it needs to be in place before technical requirements generation can begin.

Common themes from our informational discussions with programs using MBSE included that these classification requirements are difficult to deal with properly (given the self-contained nature of MBSE models) and that only a handful of computers were properly set up to run MBSE software, creating a potential bottleneck. Implementing all models at the highest classification level and implementing need-to-know procedures rather than building multiple parallel models at different classification levels seems to be the best approach to dealing with the classification problem. Purchasing a sufficient number of software licenses or running the MBSE software in a distributed cloud environment might mitigate the bottleneck caused by having only small numbers of properly configured computers.

Ensuring government compatibility with industry models is key to the usefulness of DAF MBSE models. This requires two separate pieces: First, contractor models and government models must be technically compatible. SysML provides a common language for different models to communicate. However, if other proprietary models are incorporated into an industry MBSE model, communication is more complicated. Second, contractors probably must be incentivized (through specific contract language) to keep their models current and comprehensive. Third, the DAF must also keep its models current and not lag industry. SMEs we spoke with mentioned the great benefits that could come from analyzing design changes the contractor makes in near real time but expressed worry that poor contracting language could mean that industry models update government models only updated infrequently or that key pieces of data will be missing. Therefore, it would be helpful for the contracting officer to consult the engineers when writing contract language to make sure that modeling requirements are considered.

Model-Based Systems Engineering Operations and Training

Second, significant training is required to become proficient in MBSE software, even for someone armed with a strong technical background at the onset. Therefore, DAF programs need to either invest in training for their engineers or hire contractors who are already trained. Although the latter may be more expedient, our research suggests that the most important benefits of digital engineering will become known only if the entire government engineering team fully embraces it, i.e., if an organic capability is established.

This second component appears to be a significant bottleneck in practice. Despite efforts to incorporate MBSE in a variety of PMOs, we learned that many programs lacked the personnel to take full advantage of MBSE. In some cases, only a handful of individuals had training with the appropriate software and SysML.

PMOs should therefore ensure that MBSE training is incorporated into plans to transition to a digital engineering paradigm; doing so will allow the PMO to analyze models produced by industry partners and will establish an MBSE-friendly culture.

We caution that these investments in technical capabilities are necessary, but not sufficient, to take full advantage of the MBSE paradigm. PMOs often view MBSE as a shortcut to a streamlined acquisition process, but this is shortsighted—sound engineering must still be done on the government side so that government reference architectures and MBSE interfaces are properly defined to begin with. Therefore, we also recommend increased investments in hiring and retaining systems engineers to work with the rest of the engineering team to ensure that their needs are being fully met by the contractor-delivered MBSE models. In addition, systems engineers can exploit MBSE models to aid in tradespace decisions (e.g., given a limited budget, where should investments be prioritized to improve safety, reliability, or some other quantity?). MBSE models can aid in making these decisions because they enable fast "what-if" analyses, but interrogating the models properly requires some expertise in this area. Additionally, care should be taken when making tradespace decisions to make them at the right time. Skilled systems engineers can help here as every decision further cements the final design.

The Bottom Line

Successful MBSE use throughout a program—or throughout the DAF—will likely require more than a few targeted investments in personnel and computers. Instead, a broader cultural shift toward digital engineering seems necessary. MBSE models, by design, are meant to be used widely by multiple different engineers working on different system elements. Incorporating MBSE widely may face resistance from veteran engineers used to having complete control over their own domains. Nevertheless, shifting a culture toward a more collaborative (rather than stovepiped) working environment is likely to bring about long-term benefits that outweigh any initial investments.

Abbreviations

AETC	Air Education Training Command
AF/A5	Deputy Chief of Staff for Strategy, Integration and Requirements
AFGSC	Air Force Global Strike Command
AFI	Air Force Instruction
AFIT	Air Force Institute of Technology
AFIT/EN	Air Force Institute of Technology Graduate School of Engineering and Management
AFIT/LS	Air Force Institute of Technology School of Systems and Logistics
AFLCMC	Air Force Life Cycle Management Center
AFLCMC/EN	Air Force Life Cycle Management Center, Engineering Directorate
AFLCMC/EZ	Air Force Life Cycle Management Center, Technical Engineering Services Division
AFLCMC/XP	Air Force Life Cycle Management Center, Plans and Programs Directorate
AFMC/ENS	Air Force Materiel Command, Systems Engineering Division
AFPD	Air Force Policy Directive
AFTC	Air Force Test Center
AOA	Analysis of Alternatives
A-STPA	Automated–Systems-Theoretic Process Analysis
CAIRIS	Computer Aided Integration of Requirements and Information Security
CDD	capability development document
CONEMP	concept of employment
CONOP	concept of operation
CPD	capability production document
COTS	commercial-off-the-shelf
DAF	U.S. Department of the Air Force
DAFPAM	Department of the Air Force Pamphlet
DAU	Defense Acquisition University
DoD	U.S. Department of Defense
DoDAF	Department of Defense Architectural Framework
DoDI	Department of Defense Instruction
DoN	U.S. Department of the Navy
FAA	Federal Aviation Administration
FFBD	functional flow block diagram
FFP	firm-fixed-price
FMECA	failure modes effects and criticality analysis

FTA	fault-tree analysis
FY	fiscal year
GAO	U.S. Government Accountability Office
GBSD	Ground-Based Strategic Deterrent
GBTS	Ground-Based Training System
ICBM	intercontinental ballistic missile
ICD	initial capabilities document
IDEF0	Integrated Definition Method
INCOSE	International Council on Systems Engineering
ISO	International Organization for Standardization
JROC	Joint Requirements Oversight Council
LVC	live, virtual, and constructive
MBSE	model-based systems engineering
MDAP	major development acquisition program
NAVAIR	Naval Air Systems Command
NASA	National Aeronautics and Space Administration
OV-1	Operational View 1
PMO	program management office
PSASS	MIT Partnership for Systems Approaches to Safety and Security
QAW	quality attribute workshop
RFI	request for information
RFP	request for proposal
RWG	requirements working group
SAF/AQ	Assistant Secretary of the Air Force for Acquisition, Technology and Logistics
SAF/AQL	Assistant Secretary of the Air Force for Acquisition, Technology and Logistics, Special Programs
SAF/AQP	Assistant Secretary of the Air Force for Acquisition, Technology and Logistics, Global Power Programs
SAF/AQQ	Assistant Secretary of the Air Force for Acquisition, Technology and Logistics, Global Reach Programs
SAF/AQR	Assistant Secretary of the Air Force for Acquisition, Technology and Logistics, Science Technology and Engineering
SAF/AQX	Assistant Secretary of the Air Force for Acquisition, Technology and Logistics, Acquisition Integration Leadership
SafetyHAT	Safety Hazard Analysis Tool
SAHRA	STPA Based Hazard and Risk Analysis Tool
SE	systems engineering (select figures only)
SME	subject-matter expert

SpecTRM	Specification Tools and Requirements Methodology
SPO	system program office
SRD	system requirements document
SS	system specification
STAMP	Systems Theoretic Accident Model and Processes
STPA	systems-theoretic process analysis
SysML	Systems Modeling Language
TPS	Test Pilot School
TR	technical requirement (select figures only)
TRL	technology readiness level
UML	Universal Modeling Language
XSTAMPP	eXtensible STAMP Platform

References

Abdulkhaleq, Asim, and Stefan Wagner, "A Controlled Experiment for the Empirical Evaluation of Safety Analysis Techniques for Safety-Critical Software," *Proceedings of the 19th International Conference on Evaluation and Assessment in Software Engineering*, New York: Association for Computing Machinery, April 2015.

AFI—*See* Air Force Instruction.

AFPD—*See* Air Force Policy Directive.

Air Force Instruction 10-601, *Operational Capability Requirements Development*, November 6, 2013. As of April 4, 2022: https://static.e-publishing.af.mil/production/1/af_a3_5/publication/afi10-601/afi10-601.pdf

Air Force Instruction 63-101/20-01, *Integrated Life Cycle Management*, Washington, D.C.: Department of the Air Force, June 30, 2020. As of August 30, 2021: https://static.e-publishing.af.mil/production/1/saf_aq/publication/afi63-101_20-101/afi63-101_20-101.pdf

Air Force Policy Directive 63-1, I*ntegrated Life Cycle Management*, Washington, D.C.: Department of the Air Force, August 7, 2018. As of August 30, 2021: https://static.e-publishing.af.mil/production/1/saf_aq/publication/afpd63-1/afpd63-1.pdf

Albon, Courtney, "T-7 Trainer Production Decision Delayed Seven Months," Inside Defense, June 18, 2021.

Asplund, Fredrik, Jad El-khoury, and Martin Törngren, "Safety-Guided Design Through System-Theoretic Process Analysis, Benefits and Difficulties," DiVA: Digitala Vetenskapliga Arkivet, September 27, 2012. As of September 3, 2021: https://www.diva-portal.org/smash/record.jsf?pid=diva2%3A556095&dswid=2219

Barbacci, Mario R., Robert Ellison, Anthony J. Lattanze, Judith A. Stafford, Charles B. Weinstock, and William G. Wood, *Quality Attribute Workshops (QAWs)*, 3rd ed., Pittsburgh, Pa.: Carnegie Mellon Software Engineering Institute, August 2003. As of August 25, 2021: https://resources.sei.cmu.edu/asset_files/TechnicalReport/2003_005_001_14249.pdf

"Boeing Commences T-7A Red Hawk GBTS Production for USAF," Airforce Technology website, December 2, 2020. As of September 3, 2021: https://www.airforce-technology.com/news/boeing-commences-t-7a-red-hawk-gbts-production-for-usaf/

Cockburn, Alistair, *Writing Effective Use Cases*, Boston: Addison-Wesley, 2001.

Cyber Resiliency Office for Weapon Systems, *Weapon System Program Protection/Systems Security Engineering Guidebook*, Vers. 2.0, March 12, 2020.

DAFPAM—*See* Department of the Air Force Pamphlet.

Dakwat, Alheri Longji, and Emilia Villani, "System Safety Assessment Based on STPA and Model Checking," *Safety Science*, Vol. 109, November 2018, pp. 130–143.

DAU—*See* Defense Acquisition University.

de Dianous, Valérie, and Cécile Fiévez, "ARAMIS Project: A More Explicit Demonstration of Risk Control Through the Use of Bow-Tie Diagrams and the Evaluation of Safety Barrier Performance," *Journal of Hazardous Materials,* Vol. 130, No. 3, March 31, 2006, pp. 220–233. As of September 3, 2021:
https://www.doi.org/10.1016/j.jhazmat.2005.07.010

Defense Acquisition University, "iCatalog Home Page," Fort Belvoir, Va., undated. As of July 29, 2021:
https://icatalog.dau.edu/onlinecatalog/careerlvl.aspx

Defense Acquisition University, *Systems Engineering Fundamentals*, Fort Belvoir, Va.: Defense Acquisition University Press, 2001. As of August 25, 2021:
https://ocw.mit.edu/courses/aeronautics-and-astronautics/16-885j-aircraft-systems-engineering-fall-2005/readings/sefguide_01_01.pdf

Defense Acquisition University, *Defense Acquisition Guidebook*, Fort Belvoir, Va., September 22, 2020. As of August 25, 2021:
https://www.dau.edu/tools/t/Defense-Acquisition-Guidebook

Delligatti, Lenny, *SysML Distilled: A Brief Guide to the Systems Modeling Language*, Boston: Addison-Wesley, 2014.

Department of the Air Force Pamphlet 63-128, *Integrated Life Cycle Management*, Washington, D.C.: Department of the Air Force, February 3, 2021. As of August 30, 2021:
https://static.e-publishing.af.mil/production/1/saf_aq/publication/dafpam63-128/dafpam63-128.pdf

Department of Defense Directive 5000.01, *The Defense Acquisition System*, Washington, D.C.: Office of the Under Secretary of Defense for Acquisition and Sustainment, September 9, 2020. As of April 4, 2022:
https://www.esd.whs.mil/Portals/54/Documents/DD/issuances/dodd/500001p.pdf

Department of Defense Handbook MIL-STD-499B, Air Force System Engineering Assessment Model, draft, undated, Not available to the general public.

Department of Defense Handbook MIL-HDBK-520A, *System Requirements Document Guidance*, Washington, D.C.: Department of Defense, December 19, 2011. As of June 3, 2022: http://everyspec.com/MIL-HDBK/MIL-HDBK-0500-0599/MIL-HDBK-520A_39996/

Department of Defense Instruction 5000.02, *Operation of the Adaptive Acquisition Framework*, Washington, D.C.: U.S. Department of Defense, January 23, 2020. As of August 25, 2021: https://www.esd.whs.mil/Portals/54/Documents/DD/issuances/dodi/500002p.pdf

Department of Defense Instruction 5000.85, *Major Capability Acquisition*, Washington, D.C.: U.S. Department of Defense, August 6, 2020. As of August 25, 2021: https://www.esd.whs.mil/Portals/54/Documents/DD/issuances/dodi/500085p.pdf?ver=2020-08-06-151441-153

de Ruijter, A., and F. Guldenmund, "The Bowtie Method: A Review," *Safety Science*, Vol. 88, 2016, pp. 211–218. As of July 29, 2021: https://doi.org/10.1016/j.ssci.2016.03.001

de Souza, Fellipe Guilherme Rey, Juliana de Melo Bezerra, Celso Hirata, Pierre de Saqui-Sannes, and Ludovic Apvrille, "Combining STPA with SysML Modeling," paper presented at the 14th Annual Systems Conference SYSCON 2020, Montréal, Canada, August 2020.

DoD—*See* U.S. Department of Defense.

DoDD—*See* Department of Defense Directive.

DoDI—*See* Department of Defense Instruction.

DoN—*See* U.S. Department of the Navy.

Elm, Joseph P., "Quantifying the Effectiveness of Systems Engineering," Software Engineering Institute, Carnegie Mellon University, 2014. As of December 8, 2021: https://resources.sei.cmu.edu/asset_files/Presentation/2014_017_101_413846.pdf

Erder, Murat, and Pierre Pureur, *Continuous Architecture: Sustainable Architecture in an Agile and Cloud-Centric World*, Waltham, Mass.: Elsevier, 2016.

Everstine, Brian W., "The Grey Wolf Arrives," *Air Force Magazine*, March 1, 2020.

GAO—*See* U.S. Government Accountability Office.

Gertler, Jeremiah, *Air Force T-7A Red Hawk Trainer*, Washington, D.C.: Congressional Research Service, R44856, September 18, 2019.

Harkleroad, E. P., A. E. Vela, and J. K. Kuchar, *Review of Systems-Theoretic Process Analysis (STPA) Method and Results to Support NextGen Concept Assessment and Validation*, Cambridge, Mass.: Lincoln Laboratory, Massachusetts Institute of Technology, PR-ATC-427, October 25, 2013. As of September 3, 2021: https://archive.ll.mit.edu/mission/aviation/publications/publication-files/atc-reports/Harkleroad_2013_ATC-427.pdf

Honour, Eric C., "Understanding the Value of Systems Engineering," *INCOSE Annual International Symposium Proceedings*, Vol. 14, No. 1, June 2004, pp. 1207–1222.

INCOSE—*See* International Council on Systems Engineering.

International Council on Systems Engineering, *Systems Engineering Handbook: A Guide for System Life Cycle Processes and Activities*, 4th ed., Hoboken, N.J.: John Wiley and Sons, Inc., 2015.

International Council on Systems Engineering, *Guide for Writing Requirements*, INCOSE-TP-2010-006-02, June 30, 2017. As of December 8, 2021: https://moodle.insa-toulouse.fr/pluginfile.php/25479/mod_resource/content/1/2017%20INCOSE%20Guide%20for%20Writing%20Reqs%20-%202017%20Update.pdf

International Organization for Standardization, *Systems and Software Engineering—System Life Cycle Processes*, standard, Geneva, Switzerland, ISO/IEC/IEEE 15288, May 2015. As of August 25, 2021: https://www.iso.org/standard/63711.html

ISO—*See* International Organization for Standardization.

King, Samuel, Jr., "Grey Wolf Begins Testing," press release, Robins Air Force Base, Georgia, February 19, 2020. As of August 30, 2021: https://www.robins.af.mil/News/Article-Display/Article/2087402/grey-wolf-begins-testing

Knowledge Based Systems, Inc., "IDEF0 Function Modeling Method," webpage, undated. As of April 7, 2022: https://www.idef.com/idefo-function_modeling_method/

Leffingwell, Dean, *Agile Software Requirements, Lean Requirements Practices, for Teams, Programs and the Enterprise*, Boston: Addison-Wesley, 2011.

Leveson, Nancy G., *Engineering a Safer World: Systems Thinking Applied to Safety*, Cambridge, Mass.: The MIT Press, 2012.

Leveson, Nancy, "Improving the Standard Risk Matrix: Part 1," Cambridge, Mass.: Department of Aeronautics and Astronautics, MIT, 2019. As of September 3, 2021: http://sunnyday.mit.edu/Risk-Matrix.pdf

Leveson, Nancy G., and John P. Thomas, *STPA Handbook*, Boston, Mass.: MIT Partnership for Systems Approaches to Safety and Security (PSASS), 2018. As of August 25, 2021: https://psas.scripts.mit.edu/home/get_file.php?name=STPA_handbook.pdf

Lightsey, Bob, *Systems Engineering Fundamentals*, Fort Belvoir, Va.: Defense Acquisition University Press, 2001.

Lorell, Mark A., Robert S. Leonard, and Abby Doll, *Extreme Cost Growth: Themes from Six U.S. Air Force Major Defense Acquisition Programs*, Santa Monica, Calif.: RAND Corporation, RR-630-AF, 2015. As of August 29, 2021: https://www.rand.org/pubs/research_reports/RR630.html

Lorell, Mark A., Leslie Adrienne Payne, and Karishma R. Mehta, *Program Characteristics That Contribute to Cost Growth: A Comparison of Air Force Major Defense Acquisition Programs*, Santa Monica, Calif.: RAND Corporation, RR-1761-AF, 2017. As of August 25, 2021: https://www.rand.org/pubs/research_reports/RR1761.html

Mayer, Lauren A., Paul DeLuca, Michael E. McMahon, Michael Bohnert, Thomas C. Whitmore, and Devin Tierney, *Sustaining U.S. Navy Non-Nuclear Surface Ship Design Capability*, Santa Monica, Calif.: RAND Corporation, RR-2766-NAVY, 2019, Not available to the general public.

Mayer, Lauren A., Don Snyder, Guy Weichenberg, Danielle C. Tarraf, Jonathan W. Welburn, Suzanne Genc, Myron Hura, and Bernard Fox, *Cyber Mission Thread Analysis: An Implementation Guide for Process Planning and Execution*, Santa Monica, Calif.: RAND Corporation, RR-3188/2-AF, 2022.

MIL-HDBK-520A—*See* Department of Defense Handbook MIL-HDBK-520A.

MIT Partnership for Systems Approaches to Safety and Security (PSASS), "STAMP Tools," webpage, undated. As of July 29, 2021: http://psas.scripts.mit.edu/home/stamp-tools/

MITRE Corporation, *Systems Engineering Guide: Collected Wisdom from MITRE's Systems Engineering Experts*, Bedford, Mass., 2014.

Montes, Daniel R., *Using STPA to Inform Developmental Product Testing*, dissertation, Cambridge, Mass.: Massachusetts Institute of Technology, 2016. As of August 19, 2021: https://dspace.mit.edu/handle/1721.1/103422

NASA—*See* National Aeronautics and Space Administration.

National Aeronautics and Space Administration, *Expanded Guidance for NASA Systems Engineering*, Vol. 1: *Systems Engineering Practices*, Washington, D.C., 2016a. As of April 7, 2022:
https://ntrs.nasa.gov/citations/20170007238

———, *Expanded Guidance for NASA Systems Engineering*, Vol. 2: *Crosscutting Topics, Special Topics, and Appendices*, Washington, D.C., 2016b. As of April 7, 2022:
https://ntrs.nasa.gov/citations/20170007239

———, *NASA Systems Engineering Handbook*, Washington, D.C., SP-2016-6105, 2020. As of June 3, 2022:
https://www.nasa.gov/sites/default/files/atoms/files/
nasa_systems_engineering_handbook_0.pdf

National Research Council, *Pre-Milestone A and Early-Phase Systems Engineering: A Retrospective Review and Benefits for Future Air Force Systems Acquisition*, Washington, D.C.: National Academies Press, 2008.

NAVAIR—*See* Naval Air Systems Command.

Naval Air Systems Command 4355.19E, *Systems Engineering Technical Review Process*, Patuxent River, Md., February 6, 2015.

Naval Systems Engineering Steering Group, *Naval Systems Engineering Guide*, Washington, D.C., 2004. As of August 16, 2021:
https://apps.dtic.mil/sti/citations/ADA527494

Nicholds, Boyd, Cornelis Bil, Pier Marzocca, John Mo, Murray Stimson, and David Holmes, "Case Study of Quality Function Deployment Method in Defense Missions," 2018 AIAA Aerospace Sciences Meeting, Kissimmee, Fla., January 8–12, 2018. As of April 7, 2022:
https://arc.aiaa.org/doi/10.2514/6.2018-1753

Office of the Under Secretary of Defense for Acquisition and Technology, *Systems Engineering Guide for Systems of Systems*, Washington, D.C., August 2008.

Phipps, Ron, "Writing Quality Requirements Statements," Wright-Patterson Air Force Base, Ohio: Air Force Institute of Technology, 2021.

Polidore, Kerri, "Value of Systems Engineering," presented at the 13th Annual Systems Engineering Conference, San Diego, Calif., October 27, 2010. As of December 8, 2021:
https://ndia.dtic.mil/2010/2010systemengr.html

PSASS—*See* MIT Partnership for Systems Approaches to Safety and Security.

Pub. L. 101-510, National Defense Authorization Act for Fiscal Year 1991, November 5, 1990. As of April 7, 2022:
https://www.govinfo.gov/content/pkg/STATUTE-104/pdf/STATUTE-104-Pg1485.pdf

Redshaw, Mary C., "Building on a Legacy: Renewed Focus on Systems Engineering in Defense Acquisition," *Defense AR Journal*, Vol. 17, No. 1, 2010, pp. 93–110.

Reilly, Brianna, "UH-1N Huey Replacement Delays Free in Funds in USAF Reprogramming Request," *Inside Defense*, July 8, 2021.

SAF/AQ, *Early Systems Engineering Guidebook*, version 1, Washington, D.C.: U.S. Department of the Air Force, March 31, 2009.

Scarinci, Andrea, Amanda Quilici, Danilo Ribeiro, Felipe Oliveira, Daniel Patrick and Nancy G. Leveson, "Requirement Generation for Highly Integrated Aircraft Systems Through STPA: An Application," *Journal of Aerospace Information Systems*, Vol. 16, No. 1, January 2019. As of April 7, 2022:
https://arc.aiaa.org/doi/10.2514/1.I010602

Schwartz, Moshe, *How DOD Acquires Weapon Systems and Recent Efforts to Reform the Process*, Washington, D.C.: Congressional Research Service, RL-34026, May 23, 2014. As of June 3, 2022:
https://crsreports.congress.gov/product/pdf/RL/RL34026

SD-5, *Market Research: Gathering Information About Commercial Products and Services*, Fort Belvoir, Va.: Defense Standardization Program, August 24, 2008. As of June 3, 2022:
http://everyspec.com/DoD/DoD-PUBLICATIONS/SD-5_2008_38289/

SD-15, *Guide for Performance Specifications*, Fort Belvoir, Va.: Defense Standardization Program, August 24, 2009. As of June 3, 2022:
http://everyspec.com/DoD/DoD-PUBLICATIONS/SD-15_24AUG2009_25067/

Secretary of the Air Force Public Affairs, "Air Force Announces Newest Red Tail: 'T-7A Red Hawk,'" press release, September 16, 2019. As of September 3, 2021:
https://www.af.mil/News/Article-Display/Article/1960964/air-force-announces-newest-red-tail-t-7a-red-hawk/

Shamieh, Cathleen, *System Engineering for Dummies*, IBM Limited Edition, Hoboken, N.J.: Wiley Publishing, Inc., 2011.

Skujins, Rome, "EZS-107, System Requirements Document Preparation," course materials, Wright-Patterson Air Force Base, Ohio: Air Force Life Cycle Management Center, January 28, 2021.

Snyder, Don, Elizabeth Bodine-Baron, Dahlia Goldfeld, Bernard Fox, Myron Hura, Mahyar A. Amouzegar, and Lauren Kendrick, *Cyber Mission Thread Analysis: A Prototype Framework for Assessing Impact to Missions from Cyber Attacks to Weapon Systems*, Santa Monica, Calif.: RAND Corporation, RR-3188/1-AF, 2022. As of April 7, 2022: https://www.rand.org/pubs/research_reports/RR3188z1.html

Stuckey, Richard M, Shahram Sarkani, and Thomas A. Mazzuchi, "Complex Acquisition Requirements Analysis Using a Systems Engineering Approach," *Defense AR Journal*, Vol. 24, No. 2, April 2017, pp. 266–301.

Systems Engineering Development and Implementation Center, *NAVAIR SETR Process Handbook*, v. 1.0, Patuxent River, Md.: Naval Air Systems Command, February 6, 2015.

Tirpak, John A., "Boeing Gearing Up for MH-139 Test Models Production," *Air Force Magazine*, May 20, 2019.

"USAF Rewriting Huey Replacement RFP," webpage, Rotor and Wing International, March 6, 2017. As of August 30, 2021: https://www.rotorandwing.com/2017/03/06/usaf-rewriting-huey-replacement-rfp

U.S. Department of Defense, *2015 Selected Acquisition Report (SAR): Global Positioning System III (GPS III)*, March 23, 2016. As of June 2, 2022: https://www.esd.whs.mil/Portals/54/Documents/FOID/Reading%20Room/Selected_Acquisition_Reports/FY_2015_SARS/16-F-0402_DOC_20_GPS_III_DEC_2015_SAR.pdf

U.S. Department of Defense, "Quick Search: ASSIST," webpage, 2018, September 2, 2021. As of September 3, 2021: https://quicksearch.dla.mil/qsDocDetails.aspx?ident_number=106786

U.S. Department of Defense Chief Information Officer, "DoD Architecture Framework Version 2.02, DoD Deputy Chief Information Officer," webpage, 2010. As of August 25, 2021: https://dodcio.defense.gov/Library/DoD-Architecture-Framework/dodaf20_ov1/

U.S. Department of the Navy, *Naval Systems Engineering Guide*, Washington, D.C.: Naval Systems Engineering Steering Group, October 2004. As of July 28, 2021: https://apps.dtic.mil/sti/pdfs/ADA527494.pdf

U.S. Government Accountability Office, *Weapons Acquisition Reform: Actions Needed to Address Systems Engineering and Developmental Testing Challenges*, GAO-11-806, Washington, D.C., September 2011.

U.S. Government Accountability Office, *Defense Acquisitions: Assessments of Selected Weapon Programs*, GAO-13-294SP, Washington, D.C., March 28, 2013.

U.S. Government Accountability Office, *Acquisition Process: Military Service Chiefs' Concerns Reflect Need to Better Define Requirements before Programs Start*, GAO-15-469, Washington, D.C., June 2015. As of August 25, 2021: https://www.gao.gov/assets/gao-15-469.pdf

U.S. Government Accountability Office, *Weapon System Requirements: Detailed Systems Engineering Prior to Product Development Positions Programs for Success*, GAO-17-77, Washington, D.C., November 2016. As of August 25, 2021: https://www.gao.gov/assets/gao-17-77.pdf

U.S. Government Accountability Office, *Report to Congressional Committees: Weapon Systems Annual Assessment*, Washington, D.C., GAO-21-222, June 2021. As of August 25, 2021: https://www.gao.gov/assets/gao-21-222.pdf

U.S. Office of Personnel Management, *Air Force Occupational Analysis for the Systems Engineer Position*, Washington, D.C., 2016.

Wasek, James, Shahram Sarkani, and Thomas Mazzuchi, "Measuring Defense Acquisition Capabilities with QFD," *Military Operations Research*, Vol. 14, No. 2, 2009, pp. 75–92.

Whitcomb, Clifford, Rabia Khan, and Corina White, *Development of a System Engineering Competency Career Development Model: An Analytical Approach Using Blooms Taxonomy*, Monterey, Calif.: Naval Postgraduate School, 2014.

Whitehead, N. Peter, *The Dimensions of Systems Thinking—An Approach for a Standard Language of Systems Thinking*, dissertation, Charlottesville, Va.: Department of Systems and Information Engineering, University of Virginia, 2014. As of September 3, 2021: https://libraetd.lib.virginia.edu/public_view/3x816m893

Yoo, Sam M., Andrew N. Kopeikin, Dro J. Gregorian, Adam T. Munekata, John P. Thomas, and Nancy G. Leveson, "System-Theoretic Requirements Definition for Human Interactions on Future Rotary-Wing Aircraft," *97th International Symposium on Aviation Psychology*, 2021, pp. 334–339.

Young, William Edward, and Nancy G. Leveson, "Systems Thinking for Safety and Security," *Proceedings of the 29th Annual Computer Security Applications Conference (ACSAC '13)*, Association for Computing Machinery, New York, December 2013.

Young, William, and Nancy G. Leveson, "Inside Risks: An Integrated Approach to Safety and Security Based on Systems Theory," *Communications of the ACM*, Vol. 57, No. 2, February 2014.

Zhang, Zheying, "Effective Requirements Development —A Comparison of Requirements Elicitation Techniques," working paper, 2007. As of September 3, 2021: https://citeseerx.ist.psu.edu/viewdoc/download?doi=10.1.1.135.185&rep=rep1&type=pdf

CPSIA information can be obtained
at www.ICGtesting.com
Printed in the USA
BVHW020801271222
655045BV00009B/150